FOCUS ON BRAIN RESEARCH

FOCUS ON BRAIN RESEARCH

CASEY J. RESCH
EDITOR

Nova Biomedical Books
New York

For permission to use material from this book please contact us:
Telephone 631-231-7269; Fax 631-231-8175
Web Site: http://www.novapublishers.com

LIBRARY OF CONGRESS CATALOGING-IN-PUBLICATION DATA

Focus on brain research / Casey J. Resch (editor).
p. ; cm.
Includes bibliographical references and index.
ISBN 13 978-1-60021-395-3
ISBN 10 1-60021-395-2
1. Brain. 2. Neurophysiology. 3. Brain--Pathophysiolgy. I. Resch, Casey J.
[DNLM: 1. Brain--physiology. 2. Brain Diseases--physiopathology. WL 300 F6515 2006]
QP376.F63 2006
612.8--dc22

2006024659

Published by Nova Science Publishers, Inc. ✢ New York

Contents

Preface		**vii**
Chapter I	Oxidative Stress and Disruption of the Nervous Cell *Maria Alexandra Brito, Alexandra Isabel Rosa, Rui F. M. Silva, Ana Sofia Falcão, Adelaide Fernandes and Dora Brites*	**1**
Chapter II	Neuroinflammation and Impairment of Glial Cell Regulation: Implications for the Pathogenesis of Alzheimer's Disease *Rommy von Bernhardi*	**35**
Chapter III	The Blood Brain Barrier: An Introduction with an Historical Overview *Eve M. Taylor*	**87**
Chapter IV	The Assessment of Functional Brain Lateralization: Methodological Aspects *Andreas Jansen and Stefan Knecht*	**107**
Chapter V	EEG-Mapping Study of Sex Differences during Visual Creative Thinking *Nina V. Volf, Olga M. Razumnikova and Irina V. Tarasova*	**123**
Chapter VI	A Soy-Free Diet Can Lower Testosterone Levels in Male Hamsters: Implications for Aggression Research *Timothy O. Moore and Mary Karom*	**143**
Index		**155**

PREFACE

The brain is omnipresent in bodily functions. Its vast array of responsibilities includes the regulation of all physical movements, memory, learning and emotions, as well as the reception and interpretation of all sensory inputs. These numerous responsibilities require the facility of a very complex organ, and the brain is complex, to say the least. The brain's complexity provides a constant challenge for scientists to explore new theories. Because of the brain's wide-ranging role in bodily functions these studies have an impact on many other realms of science, such as psychology, neurology, and genetics. This new book features recent studies from this important field of scientific research.

Chapter 1 - Oxidative stress results from the unbalance between reactive oxygen species production and their detoxification, and has been implicated in the cell death pathways of several brain disorders. The present article summarizes some concepts related to oxidative stress in the central nervous system and how these events take place in different neurological diseases. Possible therapeutic approaches to some of the cases are also summarized. Most studies reviewed herein refer to the brain as a particularly sensitive organ to oxidative damage, due to its high metabolic rate and also because differentiated neurons cannot divide and replace dead cells. Each brain cell type, with its unique configuration and function, responds differently to oxidative insult. Oxidative stress has been demonstrated to be involved in Parkinson's disease, Alzheimer's disease, stroke, prion diseases and bilirubin encephalopathy. The study and comprehension of the cell death pathways resulting from oxidative damage has pointed out specific therapeutic targets allowing, for instance, the usage of antioxidant drugs in animal models of neurodegenerative disorders, some of which have reached clinical trials. This has brought some hope in the battle against such disabling diseases.

Chapter 2 - Inflammation is linked to neuronal cell death in several neurodegenerative diseases including Alzheimer's disease (AD). This association is based on neuropathological studies and the ability of anti-inflammatory compounds to modify the incidence of AD. Whether inflammation is a primary cause or only one of the consequences of the neurodegenerative process is still unclear. However, a growing body of evidence implicates inflammation in at least the progression of cellular injury that leads to neuropathology in AD. The author proposes that inflammation and cellular stresses associated with aging are primary initiating events in the development of AD. A dysregulated inflammatory response can elicit glial cell activation by compounds that are poorly reactive under normal conditions. Inflammation can also be the major cause of the defective handling of beta-amyloid (Aβ) and its precursor, amyloid precursor protein (APP), by glial cells, resulting in the accumulation and aggregation of Aβ and pathological pro-inflammatory activity. Glial cell activation in response to pro-inflammatory cytokines (which are up-regulated by many conditions including systemic diseases and local injury of the central nervous system) involves a complex range of responses that include morphologic, functional, and metabolic changes. Moreover there is an increased expression of genes involved in both nitric oxide (NO) and cytokine synthesis, agents that can cause neurodegeneration. Research has shown that microglial cell responses to APP and Aβ are low compared with those to pro-inflammatory cytokines and other mediators. However, activation is enhanced under pro-inflammatory conditions, indicating that glial cell responses to Aβ related proteins can be critically dependent on the priming of glial cells by pro-inflammatory factors. Here the author reviews the evidence for inflammation's involvement in Aβ induced neurodegeneration. This evidence supports the theory that damage secondary to inflammation is not only a primary cause of neurodegeneration but also an initiator of the accumulation of Aβ in AD. The author proposes that the accumulation of Aβ, oxidative and nitrosative stresses, synaptic dysfunction and eventually neurodegeneration are related to the inflammatory status which itself is the primary event triggering dysregulation of glial cell activation. The responses of dysfunctional glia result in impaired neuronal function in AD, as well as in many slowly progressive neurodegenerative disorders.

Chapter 3 - The blood-brain barrier (BBB) creates and maintains the highly regulated extracellular environment of the CNS through three "lines of defense": - (1) a physical barrier formed by tight junctions between endothelial cells of the brain capillaries and epithelial cells of both the choroid plexus and the arachnoid membrane (Figure 1), (2) transporters that mediate the efflux of compounds from brain to blood, and (3) an enzymatic barrier conferred by an enrichment of degradative enzymes in these endothelial and epithelial cells. A perusal of 20th and 21st century literature on the BBB quickly reveals that the major focus of BBB research has been the first of these components; this is particularly highlighted by the fact that much of the effort to develop drugs that can effectively distribute into the CNS has been based on conquering this component of the barrier. While several drug delivery strategies have been developed including the osmotic opening of the BBB and vector-based transport mechanisms (Madrid et al., 1991; Pardridge, 1998, 2002), optimization of the physicochemical characteristics of drugs has been the primary focus of the pharmaceutical industry when "designing" drugs which will penetrate the BBB; indeed the octanol:water partition coefficient (log P) is often used as a primary screen in the drug discovery process (Li and Kerns, 2003). However, upon examination of the correlation between lipophilicity and BBB penetration it is apparent that there are several compounds for which the correlation between these two measures is poor and they do not penetrate the brain to the level predicted by logP (Banks, 1985; Pardridge, 1998; Golden and Pollack, 2003). In many cases this is likely because these compounds are transported from brain to blood by efflux transporters. The goal of this book is to provide a comprehensive review of the information available on efflux transport across the blood-brain barrier and to examine the role of this process in the preservation of CNS homeostasis with respect to both endogenous and exogenous substances.

Chapter 4 - The brain areas subserving cognitive functions frequently show an unequal involvement across hemispheres. The left hemisphere's superiority for language and the right hemisphere's superiority for tasks like attentional monitoring are the most obvious of these functional lateralizations. The investigation of hemispheric lateralization is not only of theoretical interest, but also relevant in the clinical context. For instance, the degree of language lateralization is a marker for the susceptibility to unilateral brain lesions, and a lack of hemispheric control due to decreased lateralization is discussed as one of the mechanisms contributing to various neuropsychiatric disorders, like autism, attention deficit hyperactivity disorder or schizophrenia.

This chapter is divided into two sections. In the first part, the authors present a short overview on the functional lateralization of cognitive brain functions and discuss applications and limitations of laterality research in various clinical and non-clinical domains. In the second section the authors outline some methodological aspects that have to be addressed in fMRI laterality studies to deliver meaningful results. Special emphasis is given to the question how - in terms of reliability and information content - the lateralization of a cognitive function can be characterized.

Chapter 5 - The study investigated sex-related differences in cortical activity during performance of a figural creative task: generating visual images on the base of a presented fragment. Additionally, the authors were interested in whether there are sex differences in the influence of instructions "to create images" (i) or "to create unique images" (ui) on the creativity scores and EEG-patterns. The study is based on 26 students (13 males and 13 females). EEG was recorded from 16 sites. The spectral power density and EEG coherence were computed for six frequency ranges between 4-30 Hz. Under (ui) vs (i) instruction fluency score decreased in both groups while originality scores insignificantly increased. Sex differences between power changes during imagery task were found only in the theta1,2-bands. Independent of instruction, power decreased more in the right than in the left hemisphere in the female group; no hemispheric differences were found in males. Sex differences in task induced changes of coherence were most distinct in the theta and beta bands. In the theta bands, these changes manifest themselves mainly as increases of coherence in females but decreases in males. Under (i) instruction figural thinking was characterized by increased coherence over the left hemisphere and over the left frontal regions paired with right posterior regions in females in comparison to males. Under (ui) instruction sex differences were mainly presented in indices of interhemispheric cooperation. Changes in the beta frequency bands manifest themselves as more coherence increase in males in comparison with females. In the beta1, under (i) instruction males vs. females demonstrated increased coherence in the left frontal and between left and right frontal regions. The interhemispheric coupling between posterior areas in the beta1 seems to be stronger in males than in females especially under (ui) instruction. In the beta2 frequency band only in male group, diffuse intra- and interhemispheric short- and long–distance coherence increases were found under both instructions. This effect was more pronounced under (i) instruction, when task solutions were not limited by originality requirements. The results indicate that brain mechanisms engaged by men and women during creative figural thinking are different.

Chapter 6 - Standard animal lab chow is soy-based, and diets rich in soy protein contain estrogen-like compounds called phytoestrogens. The purpose of this study was to explore the hormonal effects of a soy versus a soy-free diet on testosterone levels in male hamsters. In Experiment #1, thirteen adult male hamsters (110-125 g) were individually-housed, and then divided into a soy and a soy-free group. All hamsters had continuous access to food and water, and they were sacrificed after seven weeks. Blood was collected to measure testosterone levels and the testes were removed to analyze the morphology. There was a significant difference in testosterone levels between the Soy (9.11±1.26 ng/ml) and the Soy-Free group (3.89±0.27 ng/ml)[$t(11)=3.74$; $p<0.005$]. There was no difference in testes weight, but the dimensions were different between the Soy (251.2±6.9 mm2) and the Soy-Free group (229.5±6.8 mm2)[$t(13)=2.2$; $p<0.05$]. In Experiment #2, twenty adult male hamsters (110-125 g) were individually-housed and then divided into four groups: 1) Castrate+Soy; 2) Castrate+Soy-Free; 3) Sham Castrate+Soy; and 4) Sham Castrate+Soy-Free. Animals were sacrificed seven weeks after surgery and the diet manipulation. There were no detectable levels of testosterone in castrated males. However, the Sham+Soy (12.21±2.95 ng/ml) had significantly higher levels of testosterone compared to Sham+Soy-Free (3.92±0.8 ng/ml)[$t(9)=2.95$; $p=0.01$]. In sum, the authors have demonstrated that a soy-free diet lowered endogenous levels of testosterone, and they recommend that investigators evaluate the impact of phytoestrogens when investigating behavioral experiments that involve neurosteroids.

In: Focus on Brain Research
Editor: Casey J. Resch, pp. 1-33

ISBN 1-600021-395-2

Chapter I

OXIDATIVE STRESS AND DISRUPTION OF THE NERVOUS CELL

Maria Alexandra Brito, Alexandra Isabel Rosa, Rui F. M. Silva, Ana Sofia Falcão, Adelaide Fernandes and Dora Brites

Centro de Patogénese Molecular (UBMBE), Faculdade de Farmácia,
University of Lisbon, Av. Das Forças Armadas,
1600-083 Lisbon, Portugal

ABSTRACT

Oxidative stress results from the unbalance between reactive oxygen species production and their detoxification, and has been implicated in the cell death pathways of several brain disorders. The present article summarizes some concepts related to oxidative stress in the central nervous system and how these events take place in different neurological diseases. Possible therapeutic approaches to some of the cases are also summarized. Most studies reviewed herein refer to the brain as a particularly sensitive organ to oxidative damage, due to its high metabolic rate and also because differentiated neurons cannot divide and replace dead cells. Each brain cell type, with its unique configuration and function, responds differently to oxidative insult. Oxidative stress has been demonstrated to be involved in Parkinson's disease, Alzheimer's disease, stroke, prion diseases and bilirubin encephalopathy. The study and comprehension of the cell death pathways resulting from oxidative damage has pointed out specific therapeutic targets allowing, for instance, the usage of antioxidant drugs in animal models of neurodegenerative disorders, some of which have reached clinical trials. This has brought some hope in the battle against such disabling diseases.

Keywords: oxidative stress, Parkinson's disease, Alzheimer's disease, stroke, prion diseases, bilirubin encephalopathy, nerve cells.

ABBREVIATIONS

AA, ascorbic acid;
Aβ, amyloid β-peptide;
AD, Alzheimer's disease;
APP, amyloid precursor protein;
bFGF, basic fibroblast growth factor;
CAT, catalase;
CysGly, cysteinylglycine;
Cys, cystein;
COX-2, cyclooxygenase-2;
DHA, dehydroascorbic acid;
EDTA, ethylenediaminetetracetic acid;
EGTA, ethylene glycol bis (2-aminoethyl ether) - N, N, N' N' - tetraacetic acid;
Gln, glutamine;
GLT-1, Glutamate transporter-1;
Glu, glutamate;
Gly, glycine;
GPx, glutathione peroxidase;
GR, glutathione redutase;
GSH, glutathione (reduced form);
GSSG, glutathione disulfide;
GST, glutathione S-transferase;
γGluCys, γ–glutamylcysteine;
γGT, γ-glutamyl transpeptidase;
HNE, 4-hydroxy-2, 3-nonenal;
$HO^{\cdot}$, hydroxyl radical;
H_2O_2, hydrogen peroxide;
IKK, IκB kinase;
αKG, α-ketoglutarate;
αKIC, α-ketoisocaproate;
$L^{\cdot}$, lipid radical;
Leu, leucine;
$LOO^{\cdot}$, peroxyl radical;
LOOH, lipid hydroperoxide;
MPP^{+}, 1-methyl-4-phenyl-pyridinium;
MPPP, 1-methyl-4-phenyl-4-propionoxypiperidine;
MPTP, 1-methyl-4-phenyl-1,2,5,6-tetrahydropyridine;
NAC, N-acetylcysteine;
NGF, nerve growth factor;
NF-κB, nuclear factor kappa B; $NO^{\cdot}$, nitric oxide;
NOS, nitric oxide synthase;
$O^{\cdot-}$, superoxide anion radical;
O_2^{2-}, peroxide anion;

8-OHdG, 8-hydroxydeoxyguanosine;
$ONOO^-$, peroxynitrite;
PD, Parkinson's disease;
PrP^c, native prion protein;
PrP^{Sc} converted prion protein;
ROS, reactive oxygen species;
SOD, superoxide dismutase;
TNF, tumor necrosis factor;
UCB, unconjugated bilirubin;
$X^{\cdot}$, radical.

1. Introduction

A free radical is defined as any chemical species containing one or more unpaired electrons occupying an atomic or molecular orbital by themselves. Based on this definition, molecular oxygen is a diradical, once it contains two unpaired electrons. Molecular oxygen can originate highly reactive oxidative molecules, generally designated by reactive oxygen species (ROS). These include superoxide anion radical ($O^{\cdot-}$), peroxide anion (O_2^{2-}), hydroxyl radical ($HO^{\cdot}$) and peroxynitrite ($ONOO^-$). Major sources of ROS include mitochondrial oxidative phosphorylation and oxygenase activity which generate $O^{\cdot-}$, easily converted to hydrogen peroxide (H_2O_2) by superoxide dismutase (SOD). In the presence of iron (Fe^{2+}) and copper (Cu^+), Fenton reactions follow, originating $HO^{\cdot}$, a potent inducer of lipid peroxidation (Figure 1). Another threat to membrane lipids is $ONOO^-$, generated as a result of the interaction between nitric oxide ($NO^{\cdot}$) and $O^{\cdot-}$ [1].

ROS result from the body's homeostatic response to the presence of molecular oxygen, an unavoidable fact to aerobic organisms. They are, nevertheless, potential mediators of cell injury and death and therefore cells tend to possess mechanisms that either expel or destroy those species. These detoxifying systems can be classified in two groups. The first one includes antioxidant enzymes such as superoxide dismutase (SOD), catalase (CAT) and glutathione peroxidase (GPx) which are essentially preventive once they decompose and/or inactivate ROS, avoiding oxidative damage. This first group also comprehends mechanisms that promote the excision or reparation of ROS-induced cellular damage, such as glutathione S-transferase (GST). The second group consists of low molecular weight compounds that act primarily against peroxyl radicals, interrupting lipid peroxidation reaction. Such compounds are α-tocopherol, ascorbic acid (AA) and glutathione (GSH) [1]. The last one is particularly relevant (Figure 1) since it also serves as substract to GPx and GST [2, 3].

The survival of the cell depends on the fragile balance between production and detoxification of ROS. Thus, if by any reason excessive ROS are produced or detoxifying mechanisms fail, homeostatic balance is impaired and oxidative stress develops [4]. This condition is characterized by intracellular elevation of ROS levels, resulting in lipid peroxidation, as well as protein and nucleic acid oxidation. Lipids are particularly vulnerable to oxidative stress due to the existence of unsaturated bonds and the high solubility of oxygen in lipid bilayers.

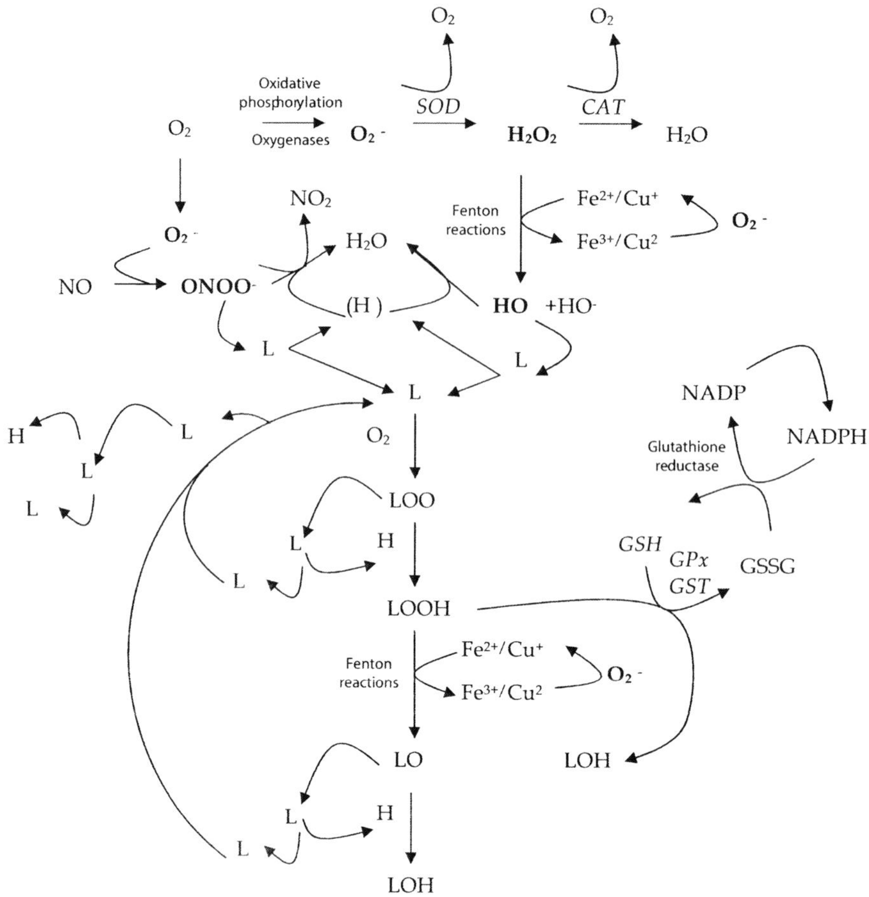

Figure 1. Production of reactive oxygen species, lipid peroxidation and major antioxidant defences. Mitochondrial oxidative phosphorylation and oxygenase activity generate superoxide ($O^{\cdot -}$), easily converted by superoxide dismutase (SOD) to hydrogen peroxide (H_2O_2). In the presence of iron (Fe^{2+}) and copper (Cu^{+}), Fenton reactions follow, leading to hydroxyl radical ($HO^{\cdot}$), a potent inducer of lipid peroxidation. Another threat to membrane lipids is peroxynitrite ($ONOO^{-}$), generated as a result of the interaction between nitric oxide ($NO^{\cdot}$) and $O^{\cdot -}$. Lipid peroxidation begins when a radical abstracts a hydrogen from a methylene group, resulting in a lipid radical ($L^{\cdot}$). This radical reacts rapidly with oxygen originating a peroxyl radical ($LOO^{\cdot}$) which can interact with other lipids to form a hydroperoxide (LOOH) and another lipid radical, perpetuating the chain reaction. Reactive oxygen species are represented in bold. Antioxidant defences such as SOD, catalase (CAT), glutathione peroxidase (GPx), glutathione S-transferase (GST) and glutathione (GSH) are represented in italic.

The rate-limiting step of lipid peroxidation begins when a radical ($X^{\cdot}$) abstracts a hydrogen from a methylene group, resulting in a lipid radical ($L^{\cdot}$). This radical reacts rapidly with oxygen originating a peroxyl radical ($LOO^{\cdot}$), which can interact with other lipids to form a hydroperoxide (LOOH) and another lipid radical, perpetuating the chain reaction (Figure 1). Peroxidation is terminated when two peroxyl radicals react to generate oxygen and a nonradical [1].

Lipid hydroperoxides can undergo rearrangement and other reactions resulting in aldehydes such as 4-hydroxy-2-nonenal (HNE), which is generated from the peroxidation of polyunsaturated fatty acids ω-6 such as arachidonic and linoleic acid. These molecules are even more toxic than, for example, $ONOO^-$ and $HO^{\cdot}$ because they have longer half-lives and can therefore diffuse to sites distant from where they were originated [5]. HNE is capable of modifying protein structure by Michael's addition, forming covalent adducts with histidine, cystein and lysine and consequently raising cellular carbonylic content. Besides the alteration in protein structure, functional impairment usually follows the formation of these adducts, resulting in inefficiency of membrane receptors and transporters, enzymes, and alterations in cell cytoskeleton, ultimately leading to cell destruction [6], as illustrated in Figure 2. Other characteristic of oxidative stress is damage of nucleic acids, DNA and RNA, which includes nucleotidic changes in sugars or bases, base deletions, strain rupture and nucleic acid-protein crosslinks. These can result in extremely serious consequences to the cell, such as mutations that lead to truncated and defective protein, inhibition of DNA, RNA and protein synthesis and cell cycle blockage [7]. Oxidative stress can ultimately lead to cell death, frequently by apoptosis [8, 9].

Compared to other tissues, the brain is especially susceptible to oxidative damage, once it consumes approximately 20% of the oxygen required by the body and it only constitutes about 2% of our body weight, indicating a high potential for generation of ROS during oxidative phosphorylation. There is also a high content of iron, which aggravates the generation of ROS, by means of Fenton reactions.

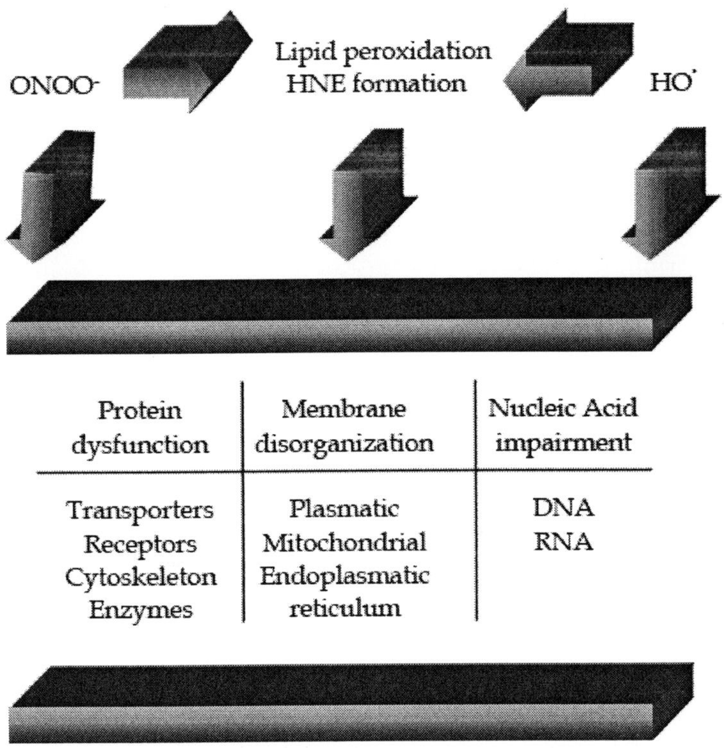

Figure 2. Simplified representation of the oxidative stress-induced effects in animal cells. Peroxynitrite (ONOO-) and hydroxyl radical (HO·) are potent lipid peroxidation initiators that, along with peroxidation products such as 4-hydroxy-2-nonenal (HNE), are capable of impairing the function and destroying cell membranes, protein and nucleic acids.

The nervous system is moreover enriched in unsaturated fatty acids, which are easy targets for lipid peroxidation. In addition, there are only low to moderate activities of antioxidant defences such as CAT, SOD and GPx and once a neuron is lost in the adult brain it cannot be replaced by generation of a new one [10].

Oxidative stress and related loss of neurons have been associated to Parkinson's disease [11, 12], Alzheimer's disease [13-17], stroke [18, 19], prionic diseases [20, 21] and bilirubin encephalopathy [22,23]. None of these conditions has a clarified pathogenesis, but only by studying the pathways of cellular damage and death in each disease will we be able to solve the puzzle of finding valuable therapeutic approaches. Thus, the objective of this article is to summarize concepts related to oxidative stress in the complex system of the human brain and how these events take place in different neurological diseases. Furthermore, some therapeutic approaches to these pathological states are also reviewed.

2. Brain Cells and Interaction between Astrocytes and Neurons

The brain is an extraordinary complex network of different cells, each one with a unique structure and function. It is this complexity that constitutes the structural basis of recognition, perception, memory, speech, locomotion and feeling, as well as hormonal and autonomic functions [24].

Neurons are the primary functional cell type of the brain with long processes called dendrites and axons. Dendrites are multiple and ramified, while axons are usually single but ramified. The interconnection of these processes allows the reception, integration and conduction of neuronal signals.

Other cell types have distinct functions: astrocytes maintain vascular, ionic and metabolic homeostasis in the mature brain. They are responsible, for instance for glutamate (Glu) uptake, sodium and potassium buffering as well as water transport. Oligodendrocytes are responsible for axon myelination, an extremely important task once neuronal signals are successfully conducted due to the existence of the highly specialized electrically insulating layer formed by myelin. Microglia, being the resident macrophage cells in the brain maintain immune surveillance [24, 25]. Recently, the possibility of the existence of a fifth nerve cell type was raised. Such cells share antigenic similarities with oligodendrocytes precursors and have been mistaken for such, even though they exist in the mature brain and are morphologically similar to astrocytes. The discovery that these cells do not depend on astrocytes to survive, and therefore respond differently than oligodendrocytes precursors to astroglial signals, lead to the hypothesis of the existence of a new type of nerve cell, named synantocyte. Synantocytes form multiple contacts with neurons, astrocytes, oligodendrocytes and myelin and apparently are involved in scar formation and protection of neural tissue [26, 27].

Different brain cells have their own functions and specialized machinery. This high level of specialization only occurs, nevertheless, because brain cells intimately interact with one another. This applies perfectly to neurons and astrocytes, whose metabolic coupling is extremely complex [28].

2.1. Glutathione Metabolism

GSH is a tripeptide (γ-L-glutamyl-L-cysteinylglycine) with several cellular functions, such as a cofactor in isomerization reactions and a storage and transport form of cystein (Cys). As a thiol, it maintains the redox potential in cells keeping sulphydryl groups of proteins in the reduced form. It is one of the major cell antioxidants as it reacts directly with radicals in nonenzimatic reactions and is also a donor of electrons in the reduction of peroxides catalyzed by GPx [2,3], as represented in Figure 3.

Reports on GSH relevance to detoxification of ROS are increasing in time [29-31]. For instance, in brain injury induced by ischemia/reperfusion, GSH levels decrease significantly while non-infarcted controls show no changes in the tripeptide levels [29]. Furthermore, inhibition of the enzymes involved in GSH synthesis results in exacerbation of brain damage [29, 30]. In addition to these functions, GSH appears to have a very special role in brain, where it has been inclusively considered a neurohormone, due to the existence of extracellular GSH in brain, the release of GSH from brain upon stimulation [32], the specific binding of GSH to extracellular receptors [32, 33], the stimulation of a signal cascade in astrocytes [34] and the induction of sodium currents in neocortex [35].

GSH is synthesized by the action of two enzymes: γ-L-glutamyl-L-cysteinylglycine (γGluCys) synthase uses Glu and cysteine (Cys) as substrates to form γGluCys, which is then combined with glycine (Gly) by GSH synthase to originate GSH (Figure 4). Both reactions require energy in the form of adenosine triphosphate (ATP) [36]. Intracellular levels of GSH are controlled by negative feedback of γGluCys synthase by GSH in order to balance synthesis and consumption. These metabolic steps apply to both neurons and astrocytes.

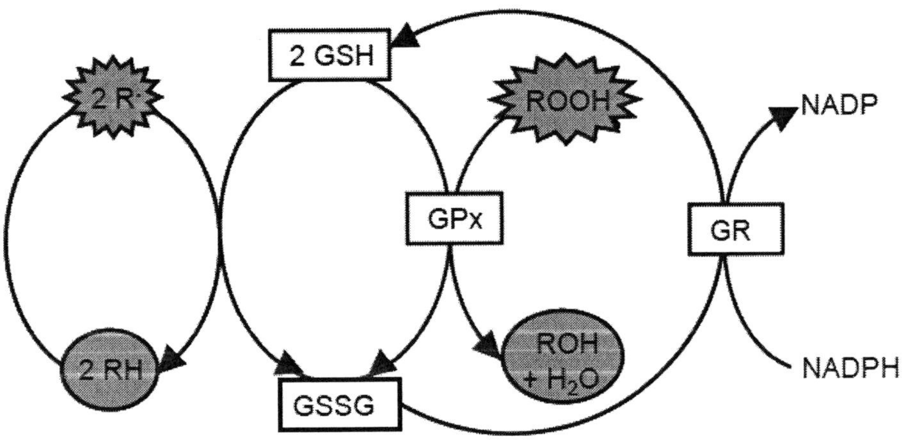

Figure 3. Schematic representation of glutathione protective role in oxidative stress. Glutathione (GSH) is one of the major cell antioxidants. It reacts directly with radicals in nonenzimatic reactions and is also a donor of electrons in the reduction of peroxides (ROOH) catalyzed by glutathione peroxidase (GPx). The resulting glutathione disulfide (GSSG) is recycled through the action of glutathione reductase (GR), in the presence of nicotinamide adenine dinucleotide phosphate (NADPH).

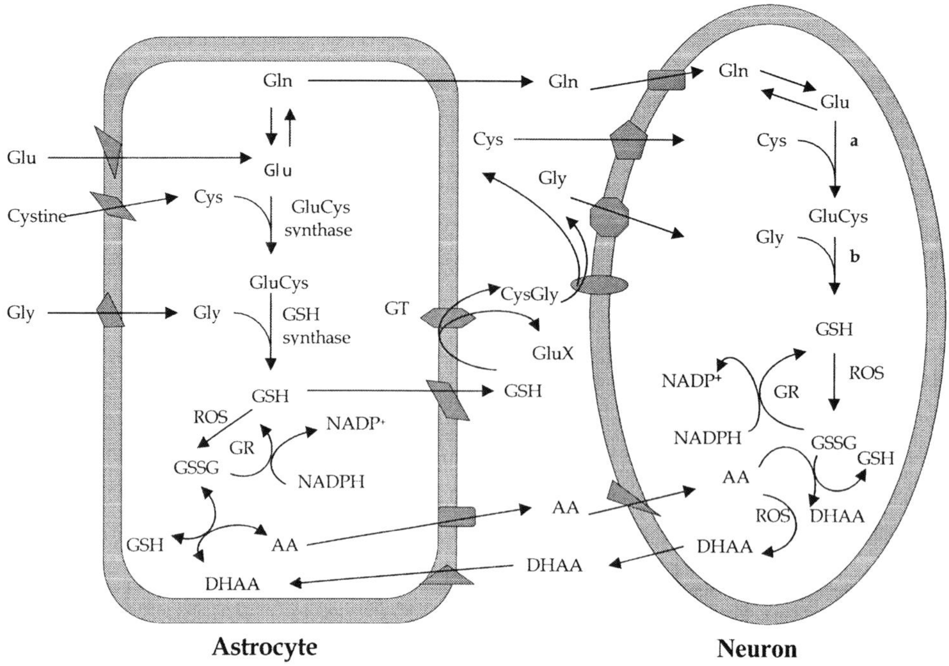

Figure 4. Schematic view of redox coupling between astrocytes and neurons. Glutathione (GSH) is synthesized in neurons and astrocytes by the action of two enzymes: γ-L-glutamyl-L-cysteinylglycine (γGluCys) synthase uses glutamate (Glu) and cysteine (Cys) as substrates to originate γGluCys, which is then combined with glycine (Gly) by GSH synthase generating GSH. Astrocytes maintain neuronal GSH by providing all three precursor amino acids. GSH is released by astrocytes and hydrolyzed by the ectoenzyme γ-glutamyl transpeptidase (γGT) originating the CysGly dipeptide, which is further hydrolysed to originate Cys and Gly, taken up for neuronal usage. Glu can be transformed to glutamine (Gln) in astrocytes, and exported to neuronal usage. While astrocytes are able to take up cystine through a sodium independent channel and break it down to yield Cys, this does not happen in neurons and therefore in these cells the rate limiting step of GSH synthesis is the usage of Cys. Reactive oxygen species (ROS) oxidise GSH to glutathione disulfide (GSSG), which is recycled back to GSH by the action of glutathione reductase (GR). Ascorbic acid (AA) is oxidized to dehydroascorbate (DHAA) in neurons due to the action of ROS or as a result of its usage for GSSG recycling. Astrocytes take up DHAA and convert it to AA using GSH, although this last reaction may be reverted.

Nevertheless, the two cell types do not compete for GSH precursors once they use different compounds to synthesise GSH. In astrocytes, the content of Glu limits GSH levels and glutamine (Gln) serves as a Glu precursor whenever it is not present. N-acetylcysteine (NAC) and, most important, cystine serve as Cys donors. The Gly pool can be derived from exogenous Gly or serine. In neuronal cultures, on the other hand, GSH levels are not limited by the availability of Glu or Gly. Cys and its precursors such as CysGly, γGluCys and NAC, however, are determinant for neurons once they cannot use cystine as a Cys donor [2]. Not only there is no competition for GSH precursors between astrocytes and neurons but there is also a strong cooperation between these cells as far as the tripeptide synthesis is concerned [37]. When both cell types are present in co-cultures, increased GSH contents are shown in neurons, compared to isolate neuron cultures [38]. This is not surprising as astrocytes indirectly contribute to neuronal GSH. In fact, astrocytes supply the most important Cys donor to neurons, CysGly: GSH is released by astrocytes and cleaved by the ectoenzyme γ-glutamyl transpeptidase (γGT), which catalyzes the transfer of the γ-glutamyl group onto an acceptor molecule to generate the dipeptide CysGly. This, in turn, is cleaved by a dipeptidase to free Cys and Gly for neuronal uptake. All three GSH precursor amino acids are then provided by astrocytes to neurons (Figure 4), especially the rate limiting Cys [37]. Some Authors have suggested the direct usage of the precursor peptides, entering the neuron via peptide transporters as described for astroglial cells [39]. The cooperation between neurons and astrocytes becomes even more interesting if one considers the role of ascorbate (AA) in GSH cycle. AA is an antioxidant molecule and it can therefore react with ROS, but it is also a cofactor for reducing GSSG. Oxidized AA, or dehydroascorbate (DHA), is released by neurons for astrocyte uptake, which in turn use GSH to convert it to AA, released for neuronal uptake and usage [37]. AA may be oxidant in the presence of transition metals, but DHA has been proven protective against oxidative damage reducing GSH depletion [40].

When GSH reacts with peroxides in the GPx reaction, it is oxidized to glutathione disulfide (GSSG). GSH is regenerated in the reaction catalysed by GR (Figs. 3 and 4), which transfers electrons from nicotinamide adenine dinucleotide phosphate (NADPH). Again, these reactions are all true for both neurons and astrocytes, with no overall consumption of GSH, which is always recycled. The same does not happen, nevertheless, with the reactions catalysed by GST, or obviously when cells release GSH, ability which has been reported in astrocytes [2]. In these cases, GSH intracellular levels decrease. To compensate for this loss, cells may use extracellular GSH and glutathione conjugates as substrates for γGT, which catalyzes the transfer of the γ-glutamyl group onto an acceptor molecule to generate the dipeptide CysGly or the CysGly conjugate. These can be hydrolysed by ectopeptidases to amino acids which can be taken up by cells [2]. Astrocytes contain greater GSH levels, as well as enzymes involved in GSH metabolism [37,41] than neurons which indicates that they are more capable of scavenging ROS. Evidence indicates that they function to support neuronal survival, an hypothesis corroborated not only by the role of astrocytes in neuronal GSH synthesis, already described, but also by the lower resistance of neuron cultures versus co-cultures of neurons and astrocytes, in the presence of $NO^{\cdot}$, H_2O_2 or $O^{\cdot-}$. Furthermore, GSH depleted astrocytes show reduced ability to protect neurons from oxidative insult [37].

2.2. Glutamate Metabolism

Neurons submitted to oxidative stress are prone to excitotoxicity [42] and therefore control of extracellular Glu levels is most important to prevent cell death. Clearance of Glu from the extracellular space is accomplished by sodium dependent transporters mainly located in astrocytes (Figure 5), maintaining Glu under the toxic threshold. The inward movement of Glu occurs against steep concentration gradient and therefore is coupled to the inward movement of three sodium ions and one hydrogen ion, and the outward movement of one potassium ion, requiring great amounts of ATP [43]. When oxidative stress occurs, ATP depletion usually follows, resulting in the ATP-supported membrane gradients collapse. This originates cessation of Glu uptake in astrocytes and also its efflux, apparently because Glu transporters, as many others, are capable of allowing transport in both inward and outward directions. Accumulation of Glu in the synaptic cleft results in excitotoxic phenomenon and neuronal death [44]. It is then clear that astrocytes play an important role in the protection of neurons against oxidative stress-induced excitotoxicity [37, 41].

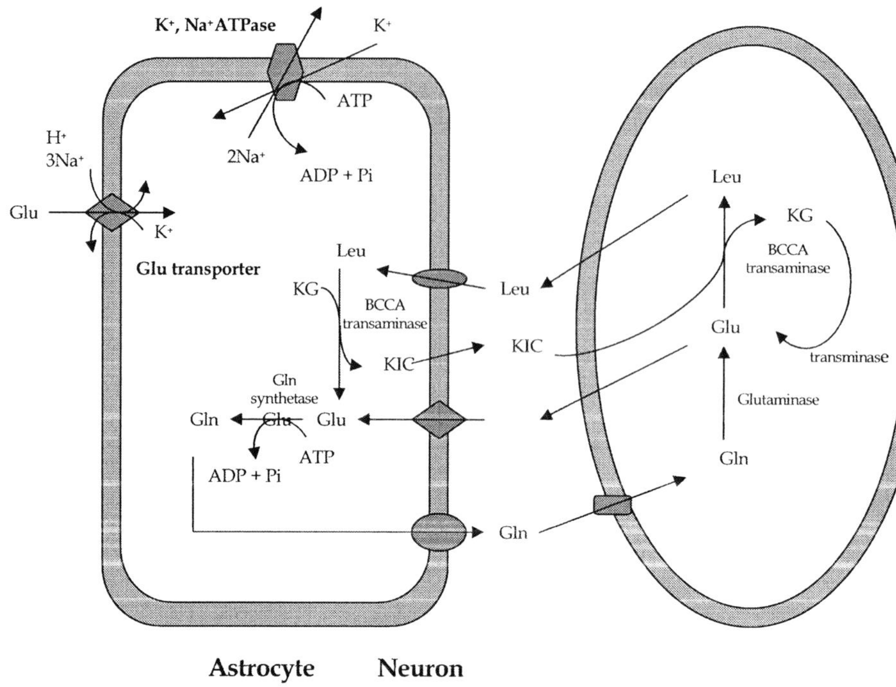

Figure 5. Simplified representation of glutamate metabolism coupling between neurons and astrocytes. Clearance of glutamate (Glu) from the extracellular space is accomplished by transporters located in astrocytes, maintaining it under the toxic threshold. The inward movement of Glu occurs against steep concentration gradient and it is therefore coupled to the inward movement of three sodium ions (Na+) and one hydrogen ion (H+), and the outward movement of one potassium ion (K+), requiring great amounts of adenosine triphosphate (ATP). Astrocytes use the Glu released by neurons which may be converted to glutamine (Gln) by Gln synthase and released in the extracellular medium for neuronal uptake. Gln is converted back to Glu in neurons by glutaminase. Neuronal Glu is also formed from α-ketoglutarate (αKG), which can result from coupling between astrocytes and neurons as well. Astrocytes take up leucine (Leu) and other branched-chain amino acids and the amino group of Leu is transferred to αKG, originating Glu and α-ketoisocaproate (αKIC). The enzyme catalysing this reaction is branched-chain amino acid (BCAA) transaminase. Thus, αKIC can then be transferred to neurons which can produce αKG through the reverse reaction.

The link between neuronal and astroglial cell types as far as Glu is concerned does not stick to its clearance and consequent neuron protection. In fact, astrocytes actually use the Glu released by neurons. An example of this is GSH synthesis, where Glu availability is critical. GSH synthesis may then be a way of disposal of toxic Glu in the brain [45]. Taken up Glu might also be converted in astrocytes to Gln by Gln synthetase and released in the extracellular medium for neuronal uptake. Gln is converted back to Glu in neurons (Figure 5), which is packaged into synaptic vesicles as any other neurotransmitter [37, 41].

Neuronal Glu is also formed from α-ketoglutarate (αKG), which can result from coupling between astrocytes and neurons as well (Figure 5). In fact, astrocytes take up leucine (Leu) and other branched-chain amino acids and the amino group of Leu is transferred to αKG originating Glu and α-ketoisocaproate (αKIC), a reaction catalysed by branched-chain amino acid transaminase. In addition, αKIC can then be transferred to neurons which can produce αKG through the reverse reaction [46].

3. Susceptibility of Different Nerve Cell Types to Oxidative Stress

The same complexity which accounts for its correct function renders the brain a particularly difficult subject to study cellular pathways and oxidative stress is no exception. Different brain cells tend to have unique responses to oxidative insults, and apparently similar cells may have opposite responses depending on the brain zone they come from [28].

3.1. The Response of Neurons towards Oxidative Stress

When oxidative stress develops in neurons, the redox balance is seriously affected, ROS are massively produced [47, 48] and, conversely, internal stores of antioxidant GSH are severely depleted [48].

On the onset of oxidative stress, ROS originate lipid peroxidation resulting in the disorganisation of cellular membranes, consequent impaired function and cell death [47,48]. ROS-attacked plasma membrane suffers fluidity changes and looses part of its selective permeability during lipid peroxidation [49]. Mitochondrial membrane permeability is also affected [50] and mitochondrial transmembrane potential collapses, leading to inhibition of electron transfer in the respiratory chain and reduced production of ATP [48], which in neurons is extremely critical since they have such high energy necessities [10]. This condition further increases production of ROS. Altered mitochondrial membrane permeability promotes cytochrome *c* release, which among other stimuli, i.e. production of ROS and collapse of mitochondrial transmembrane potential (Figure 6A) may trigger apoptosis [8, 9].

The degradation product of lipid peroxidation, HNE, may increase Ca^{2+} content indirectly by inhibition of a protein phosphatase which would augment phosphorylation of Ca^{2+}channel proteins exacerbating Ca^{2+} influx through voltage-dependent channels [51]. Excessive intracellular Ca^{2+} can kill neurons by entering the mitochondria and perturbating its function. In fact, the elevation of the mitochondrial matrix Ca^{2+} levels leads to alteration of permeability, disrupts transmembrane potential and thus energy metabolism, aggravating ROS production. The opening of a large unselective channel, called permeability transition pore, through which some caspase cofactors such as cytochrome *c* are released, results in caspase activation and apoptosis [52,53]. Ca^{2+} increases in neurons can also result in necrotic or apoptotic death due to the activation of Ca^{2+} depending proteases (calpains) [54]. Mitochondrial function impairment by HNE does not happen solely through the disruption of Ca^{2+} metabolism. In fact, caspases 3, 7 and 9 and their regulators, APAF-1 and

Smac/DIABLO (proapoptotic proteins) seem to be involved in HNE-mediated death (Figure 6A), indicating that it can also directly induce apoptosis [13].

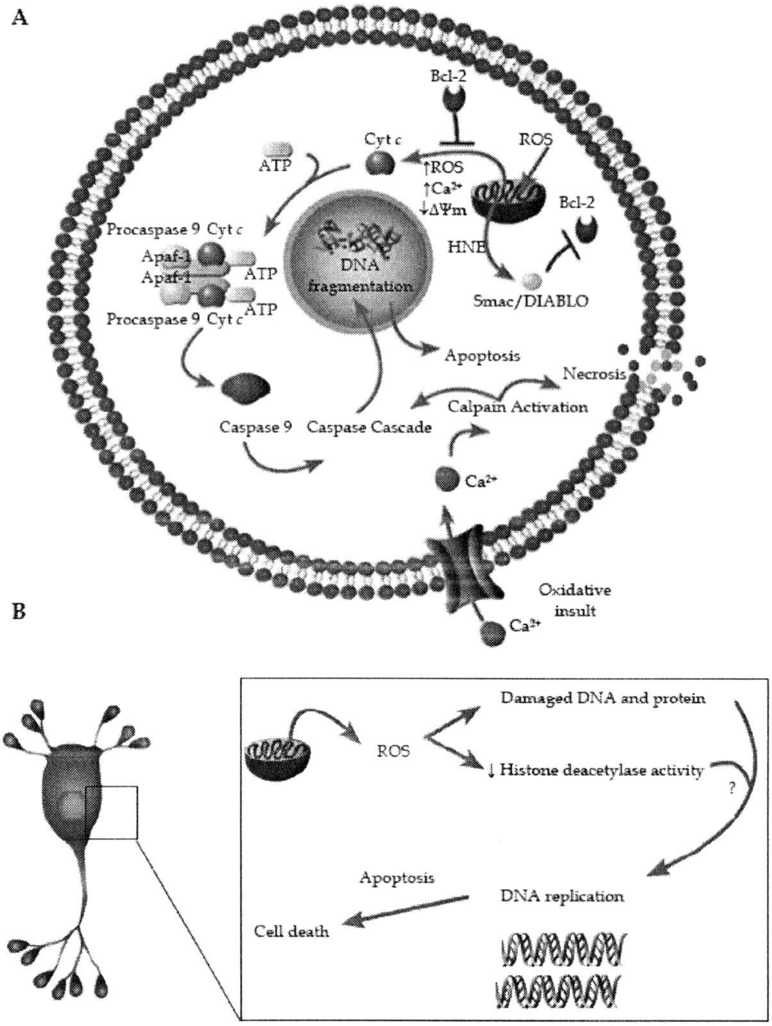

Figure 6. Mechanisms of cell death by apoptosis following exposure to oxidative stress. A. Mitochondrial membrane permeability is affected due to reactive oxygen species (ROS) action, leading to cytochrome c (Cyt c) release. Once released, Cyt c binds to apoptotic protease activating factor-1 (Apaf-1), which results in the formation of the Apaf-1-procaspase 9 apoptosome complex and activation of the initiator caspase 9. Effector caspases 3, 6 and 7 are then activated which are responsible for the cleavage of several cellular substrates originating apoptosis. Decreased transmembrane potential, as well as intracellular calcium (Ca2+) elevation, due to ROS and 4-hydroxy-2-nonenal (HNE) action, can also result in Cyt c release. Cell death may also result from the activation of Ca2+ dependent proteases, calpains. Smac/DIABLO is a proapoptotic protein once it inhibits the antiapoptotic Bcl-2 and it is released by the action of HNE. B. Cell cycle reentry in neurons might happen firstly because of ROS damage to DNA and protein and secondly due to a decrease in the histone deacetylase activity, which can result in chromatin remodelling, activation of genes, the products of which may lead to DNA replication. After reentry in cell cycle, apoptotic death follows.

HNE can also originate protein cross-linking, compromise protein function and obviously aggravate the cell stress status. Among the most common protein targets are the cytoskeleton proteins and enzymes as well as receptors and transporters [5]. In the plasma membrane, HNE apparently interacts with proteins involved in the maintenance of cellular calcium (Ca^{2+}) homeostasis such as Na^{+}/ K^{+} ATPase, Ca^{2+} ATPase, and glucose transporter, leading to an increase of intracellular Ca^{2+} following activation of Glu receptors and excitotoxicity.

In addition to HNE, $ONOO^{-}$ is thought to play an important role in the dysfunction of Ca^{2+} regulating proteins. Actually, neurons in the brains of mice lacking inducible nitric oxide synthase (NOS) showed increased resistance to excitotoxic injury, whereas treatment of cultured neurons with antioxidants such as vitamin E and GSH protected against excitoxicity [51].

During oxidative stress, DNA damage has been reported in neurons, evidenced by the presence of the modified base 8-hydroxydeoxyguanosine (8-OHdG) [55]. 8-OHdG deriving from the reaction of $HO^{\cdot}$ with purines, induces transvertions to T or C both in vitro replication and in vivo mutagenesis experiments [7]. Recent studies have surprisingly shown that oxidative stress may promote death in neurons through aberrant reentry into the cell cycle (Figure 6B). This might happen because of ROS damage to DNA and protein or due to a decrease in the histone deacetylase activity, which can result in chromatin remodelling and activation of genes [55, 56].

Oxidative damage is also capable of regulating gene expression, for antioxidant enzymes levels are frequently increased as a defence mechanism. An example of this is the upregulation of CAT and SOD in neurons [47, 48], which apparently happens through the activation of the nuclear factor κB (NF-κB) by ROS. NF-κB is a transcriptional factor which rests inactive in the cell cytosol under normal conditions as a protein complex, consisting of a repressor, IκBα, and an active dimmer [57, 58]. Signals such as elevation of intracellular ROS or Ca^{2+} and excess Glu in the synaptic cleft activate IκBα kinase (IKK), resulting in phosphorylation of IκBα . This targets the repressor for degradation in the proteasome and frees the active dimmer, which translocates to the nucleus and binds to DNA (Figure 7). The described binding augments the expression of antioxidant enzymes and prevents oxidative stress-induced apoptotic death [57]. Calbindin, a Ca^{2+} binding protein is also upregulated as a consequence of NF-κB activation. This is another obvious defensive cell response, aiming the reduction of intracellular Ca^{2+} [59]. The lipid peroxidation product HNE, however, may endanger such defensive responses of the neuron by suppressing NF-κB activation [60].

3.2. Astrocytes Response to Oxidative Stress

Exposed to oxidative stress-inducing agents, astrocytes exhibit only mild responses compared to neurons. An example of this is the ROS production induced by exposure to Paraquat, a widely used herbicide known for causing oxidative damage. In fact, a 2 h exposure to this insult originated a great increase of ROS in neurons, while in astrocytes the elevation was much less pronounced [48].

According to the same study, the cellular GSH pool was significantly larger in astrocytes than in neurons, while the depletion of the antioxidant GSH was much more pronounced in neurons than in astrocytes. Taken together, these facts comply with the differences in the production of ROS in the two types of cells and are in line with the differential susceptibility of neurons and astrocytes to oxidative insults [61].

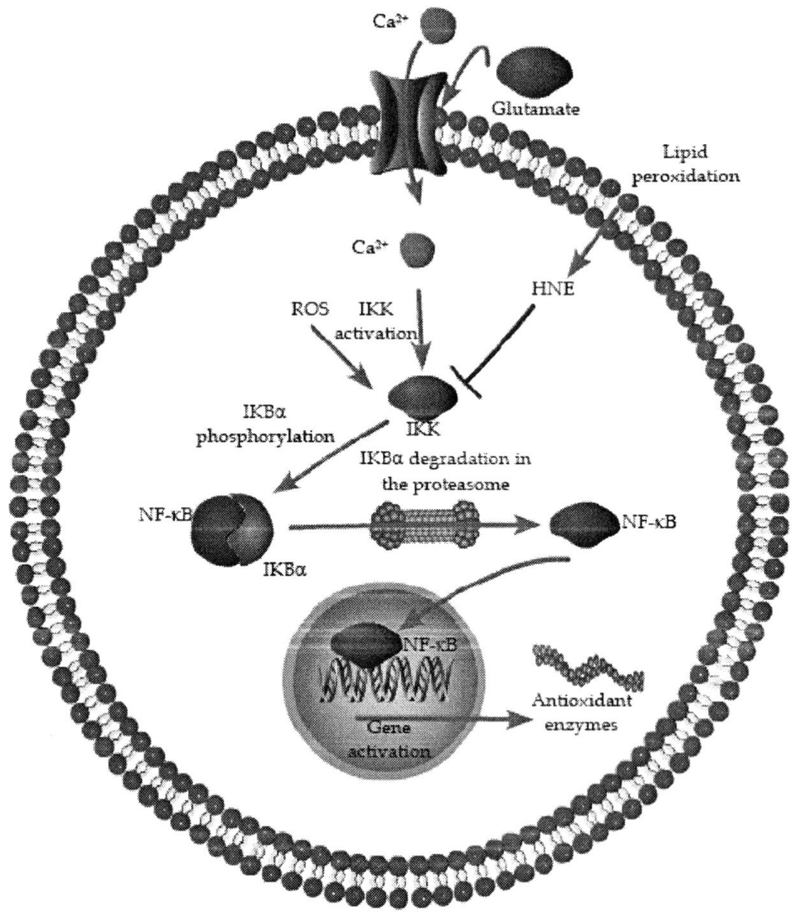

Figure 7. Simplified representation of the nuclear factor kB role in neurons exposed to oxidative stress. Nuclear factor kB (NF-kB) rests inactive in the cell cytosol bound to its inhibitor protein, IkBα. Signals such as elevation of intracellular reactive oxygen species (ROS) or Ca2+ and excess glutamate in the synaptic cleft activate IkBα kinase (IKK), resulting in phosphorylation of IkBα. This targets the repressor for degradation in the proteasome and frees the active dimmer, which translocates to the nucleus and binds to DNA, augmenting the expression of antioxidant enzymes, thus preventing apoptotic death. The lipid peroxidation product 4-hydroxy-2-nonenal (HNE), however, may endanger this defensive response of the cell by suppressing NF-kB activation.

Although the degradation product of lipid peroxidation, HNE, was encountered in astrocytes in oxidative stress damaged brain [20], it was only detected in neurons following experimental exposure of both neuronal and astroglial cells to Paraquat [48]. Therefore, it is probable that astrocytes have a mechanism for elimination of HNE, rendering them more resistant to oxidative damage.

ROS and HNE are also responsible for mitochondrial dysfunction in cortical astrocytes [62]. However, there is only a slight affection of transmembrane potential and little ATP depletion in astrocytes comparing to neurons [48,63]. Actually, astrocytes are considered "glycolytic" cells, because when exposed to mitochondrial toxins, glucose metabolism is stimulated through the glycolytic pathway, in order to maintain ATP production. Moreover, stimulation of the pentose phosphate pathway supplies NADPH, essential for GSH regeneration in the reaction catalyzed by GR [64]. Experiments have demonstrated that $ONOO^-$ inhibited cellular respiration in a time- and dose-dependent manner in neurons and astrocytes, but this resulted in stimulation of glycolysis in astrocytes and associated maintenance of ATP concentrations, on the contrary of what happened in neurons, where inhibition culminated in apoptotic death [65]. Therefore, it is understandable that some authors believe that the same degree of inhibition of mitochondrial respiration by $ONOO^-$ is proapoptotic in neurons but antiapoptotic in astrocytes [63].

Even though astrocytes are more resistant to oxidative insult, oxidised proteins are found in these cells, as evidenced by the Glu transporter-1 (GLT-1). In this regard, data suggest that HNE promotes intermolecular crosslinking of GLT-1 monomers to form dimmers, resulting in the deficiency of Glu transport by the astroglial cell and promoting excitotoxicity in surrounding neurons [62].

Oxidative stress also promotes upregulation of antioxidant enzymes in astrocytes, although not the same as in neurons. CAT and SOD levels in oxidative insult exposed astrocytes were relevantly higher when compared to non exposed cells levels [47,48]. This adaptative response to oxidative stress is, as in neurons, mediated through activation of NF-κB [51,57]. Nevertheless, in astrocytes the activation of the transcription factor results in increased expression of NOS and consequent increase in NO• production that can, in turn, damage neurons. Therefore, depending on the cell in which it is activated, NF-κB may be antiapoptotic or indirectly proapoptotic, regarding neurons [57].

3.3. Oligodendrocytes and Reactive Oxygen Species

Oligodendrocytes and their precursor cells are known for their vulnerability to oxidative stress [66] and have been identified as important targets, for instance, in stroke [67]. These cells are inclusively more sensitive to oxidative stress than neurons [68]. Apparently this happens due to the existence of risk factors such as high intracellular concentrations of iron [2, 69], which promote Fenton reactions, and because of low contents of GSH. In fact, oligodendrocytes survive in culture only when supplemented with the precursors of GSH, cysteine or cystine, otherwise the intracellular amount of the antioxidant molecule decline and cell death follows [2]. Oligodendrocyte susceptibility towards oxidative stress is critical for the nervous system as these cells are responsible for the myelinization of axons, and their death results in demyelination [2, 69]. Developing and mature oligodendrocytes show different responses to oxidative stress and this fact has helped unravelling the mechanism by which developed cells detoxify ROS such as H_2O_2. Mature oligodendrocytes dispose of H_2O_2 faster than developing cells even though GSH and CAT levels are similar in both types of oligodendrocytes. The reason underlined is the higher expression of GPx in the mature cells as, apparently, GPx is essential to protect CAT from H_2O_2 yielding greater resistance to oxidative damage [70]. When ROS production overwhelms oligodendrocytes antioxidant defences, mitochondrial damage follows, as well as cytochrome *c* release and activation of caspases 3 and 9, culminating in apoptotic death [71]. Accordingly, when ROS are produced, cell death can be prevented by free radical scavengers [2]. Synantocytes are thought to be the fifth nerve cell type and have been mistaken for oligodendrocyte precursors until recently [26]. These cells are capable of reacting to neuronal activity, glutamate and ATP, originating raised intracellular Ca2+. Generally, this results in proliferation, glial scar formation and regeneration of oligodendrocytes [27]. Since oxidative stress may promote glutamate accumulation in the synaptic cleft altering neuronal activity [51], as well as originating ATP reduction [63], it is highly likely that this novel type of nerve cell plays a central role in the response to oxidative insult.

3.4. Response of Microglial Cells to Reactive Oxygen Species

Microglial cells express the H_2O_2 reducing enzymes GPx and CAT and possess higher contents of GSH and increased specific activities of GPx and GSH reductase (GR) when compared to neurons and astrocytes. The fact comes as no surprise since microglial cells produce $O^{\cdot-}$ and $NO^{\cdot}$ on activation, which easily react with each other to originate $ONOO^{-}$. Thus, because they are in intimate contact with ROS, these cells must have potent antioxidant machinery [2].

Upon oxidative stress, microglia, as astrocytes, show increased expression of NOS and high $NO^{\cdot}$ production through the activation of NF-κB [57]. This leads to the formation of $ONOO^{-}$ that, as seen before, is toxic to neurons. Neurotoxicity may be worsened by $ONOO^{-}$ promoted injury in oligodendrocytes that may result in hypomielination of axons [66]. However, the role of microglia in neurotoxic events is not so simple, since these cells, when activated, also produce neurotrophic factors, such as nerve growth factor (NGF), basic fibroblast growth factor (bFGF) and tumor necrosis factor (TNF), which have been demonstrated to prevent neuronal death in various models of neurodegenerative disorders [57].

The neurodegenerative process is usually characterized not only by the elevation of ROS but also by inflammation. Microglia, which can be stimulated by oxidative stress, produce on activation inflammatory cytokines and ROS as by-products. Cytokines are intended to protect the organism, but overstimulation of microglia can lead to serious inflammatory and oxidative insults, resulting in cell damage. Corroborating this is the protection of neurons through the inhibition of cyclooxygenase-2 (COX-2), the rate-limiting enzyme in arachidonic acid-derived prostaglandin production, which is induced during pathological conditions such as Parkinson's disease [72].

4. Oxidative Stress and Brain Injury

Oxidative stress is a common event in neurodegenerative disorders. The barrier between cause and consequence is nevertheless frequently obscure, since the pathogenesis of the neurodegenerative process is severely complex in each disorder. It is most important, however, to enlighten the cellular pathways that result in brain cell loss, in order to find appropriate therapeutic approaches.

4.1. Parkinson's Disease

Parkinson's disease (PD) is a progressively worsening in time syndrome, usually starting on one side of the body and gradually affecting the other side. The first recognizable symptom is usually rest tremor, but sometimes the condition starts with bradikinesia (slowness and decreased amplitude of movement) and tremor may never develop. Nonmotor complaints are also described such as fatigue, depression, anxiety, sleep disturbances, constipation, bladder disturbances, sensory abnormalities, apathy, bradyphrenia (slowness in thinking) and dementia. The cause of PD is the degeneration of dopaminergic neurons in substantia nigra, resulting in deficiency of striatal dopamine concentration. This neuronal loss is accompanied by gliosis and loss of neuromelanin, a pigment contained in dopaminergic neurons [11].

Several trials have searched for the root cause of that neurodegenerative disease, raising the long known question of genetics [73, 74] against environment [12]. Up to now, aging provides the strongest correlation and men seem to show greater incidence of PD [75]. It appears that most cases are sporadic, however over 10% reveal a positive family history, and several monogenic forms of the disease have been identified [76]. Parkinsonism can nevertheless be induced artificially. This was first observed when a 23-year-old graduate student died of accidental poisoning from MPTP (1-methy-4-phenyl-1,2,5,6-tetrahydropyridine), an impurity present in the illicitly produced synthetic heroin MPPP (1-methyl-4-phenyl-4-propionoxypiperidine), where neuropathological changes, such as loss of substantia nigra neurons and neuromelanin were observed. Models using MPTP-induced parkinsonism in mice and primates have shown that its metabolite, MPP^+(1-methyl-4-phenyl-pyridinium), is specifically accumulated in dopaminergic neurons, through the dopamine uptake transporter. Once inside the neuron it depresses mitochondrial oxidative phosphorylation, resulting in ROS production. Decreasing ATP levels, a result of phosphorylative oxidation inhibition, would take its toll on the plasma membrane sodium/potassium exchanger, present in particularly high levels in dopaminergic neurons and on the ATP driven pump situated in synaptic vesicles, resulting in redistribution of dopamine in the cytosol [12]. Dopamine itself can inhibit phosphorylative oxidation [77] and initiate selective degeneration (Figure 8), even though its metabolites may produce the same effects more efficiently. These oxidant products are also capable of reacting with cellular protein including the dopamine uptake transporter itself [78].

A clue that seems to indicate that even normal substantia nigra neurons undergo some degree of oxidative stress because of dopamine is the presence of neuromelanin, specifically present in these cells. This pigment is composed mainly of dopamine-quinone, dopamine-semiquinone, as well as lipids and proteins with which the former reacted. This includes dopamine-glutathione [79]. Neuromelanin avidly binds iron and other metals which may provide a basis for high iron levels in brain, suggested to contribute to ROS formation through Fenton reactions [80]. Another reason indicating that dopamine induces neuronal death is the neuroprotection induced by the overexpression of the transporter present in synaptic vesicles, increasing dopamine in synaptic vesicles and reducing its levels in cytosol [81]. It has been told that gliosis develops in PD. Microglia particularly may participate in cell death (Figure 8) by producing noxious compounds such as pro-inflammatory cytokines, prostaglandins and ROS [82].

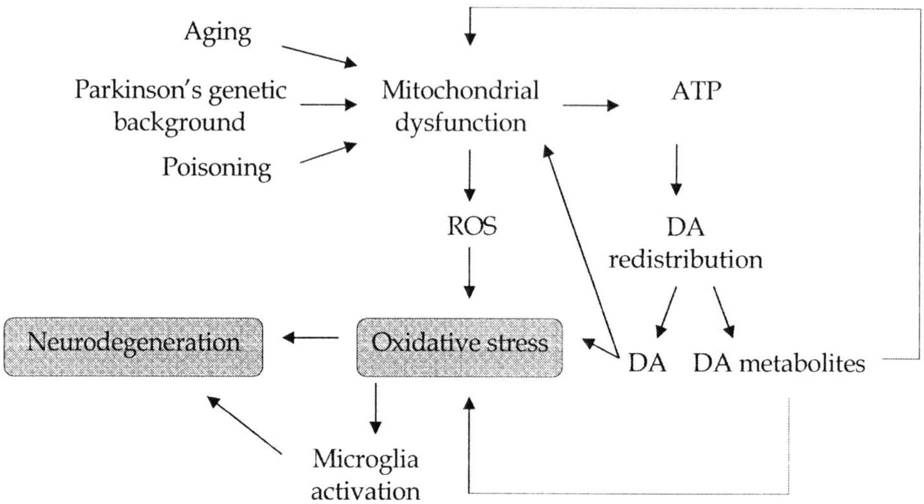

Figure 8. Representation of oxidative stress role in Parkinson's disease death pathways. Parkinson-inducing factors contribute to mitochondrial damage, resulting in reactive oxygen species (ROS) production. Decreasing ATP levels, a result of phosphorylative oxidation inhibition would take its toll on the adenosine triphosphate (ATP) driven pump situated in synaptic vesicles, resulting in redistribution of dopamine (DA) in the cytosol. Dopamine, and specially its metabolites, may initiate selective degeneration. These oxidant products are also capable of reacting with cellular protein and induce oxidative stress. Oxidative stress may also activate microglia, further contributing to neurogeneration.

Currently, there is no cure for PD. Pharmacologic approaches can only slow down the condition evolution, but stopping and reversing it is, for now, impossible. Nevertheless, new therapy concepts are being tested, specially focusing on reducing oxidative stress and consequent excitotoxicity, enhancing mitochondria function, inhibiting apoptosis and counteracting inflammation [11]. As far as oxidative stress is concerned, antioxidant supplements such as GSH and its precursor NAC seem to be neuroprotective in PD models [83].

Coenzyme Q_{10} is an antioxidant and an electron acceptor in oxidative phosphorylation existing in the precise complex, which is thought to be inhibited by MPP^+ and dopamine in parkinsonism. Tests in rats consisting of oral supplementation of this compound resulted in its higher levels in brain [84], and a clinical trial in humans has registered slowing down of the functional decline in the supplemented subjects. Investigators, nonetheless, recommended caution in the interpretation of data due to the small dimension of the trial sample [85].

Inflammation is considered to be a secondary effect of neurodegeneration, but it could also contribute to cell death through the production of cytokines, prostaglandins and ROS. As a result, COX-2 inhibitors have been tested and proven effective in the MPTP model [72].

4.2. Alzheimer's Disease

Alzheimer's disease (AD) is a commonly found in the elderly clinical condition consisting of progressive loss of memory and cognition. Pathologically it is characterized by synapse loss and formation of senile plaques, rich in the amyloid β-peptide (Aβ), originated in the cleavage of the transmembrane amyloid precursor protein (APP). Brains with this condition show serious oxidative stress and, apparently, aggregated Aβ plays a central role in originating such a state, because once inserted in nervous membranes it is able to induce ROS elevation, protein oxidation, lipid peroxidation and nucleic acid oxidation (Figure 9). Aβ is capable of inducing NOS as well, further enhancing the production of ROS [15]. Several studies have implicated imbalances of copper and iron in AD, which may aggravate oxidative stress via Fenton reactions [14].

HNE accumulation, as a consequence of lipid peroxidation, has been reported in AD [86]. HNE might explain the loss of function of the glial Glu transporter GLT-1, as the alkenal reacts by Michael addition with protein, covalently modifying its structure and function [16]. Once AD brain has a decreased Gln synthase [87, 88] activity, which is also selectively oxidized in this condition, two mechanisms of detoxifying Glu are then compromised, leading to excitotoxicity and neurodegeneration. Moreover, the activity of GST, an enzyme that detoxifies HNE, is decreased in AD, therefore aggravating HNE-induced toxicity [15]. The APOE allele is a risk for AD and it might also be associated with HNE production [89].

Several other proteins are selectively oxidized in AD and the impairment of each of them may result in cell death [17]. It is the case of creatine kinase [87,88] and α-enolase [88,90], proteins involved in ATP production, which oxidation results in diminished ATP levels. This can affect ion pumps and phospholipid asymmetry, compromising membrane function and ultimately leading to Ca^{2+}-mediated cell death. β-actin is also oxidized in AD [88], leading to abnormal cytoskeletal assembly and neurotoxicity. Ubiquitin is, in normal circumstances, added to damaged or aggregated proteins as an intracellular signal for their destruction in the proteasome. However, the cell has a reduced ubiquitin pool and so it needs to be recycled. This is the task of ubiquitin C-terminal hydrolase L-1. This enzyme is another example of selectively oxidized proteins [87,88] in AD and its alteration could result in the accumulation of damaged or aggregated proteins, including Aβ. Finally, another selectively oxidized protein in AD is dihydropyrimidinase related protein-2 [88,90], important in the elongation and directionality of neurons. Should it fail to work, and neuron contact would be compromised, which happens in AD [16, 17].

Because oxidative stress is so intimately involved in AD pathogenesis and once antioxidant activity is reduced in AD [91], treatment should comprehend antioxidant compounds administration. Examples are vitamin E or AA, and several studies have reported that increased intake of such substances can reduce the incidence of the disease [92]. This theory is controversial, however, and some authors have inclusively reported AA as a pro-oxidant due to its interaction with metal ions [93].

Chelators, capable of removing metal ions, have also proven somewhat useful in AD. For instance, ethylene glycol bis (2-aminoethyl ether) - N, N, N' N' - tetraacetic acid (EGTA) chelates metallic ions and also mimics SOD activity when complexed to such ions, reducing oxidative stress and dissolving plaques [92]. An iron-chelating agent, desferrioxamine, slowed the clinical development of the disease in preliminary clinical studies [94].

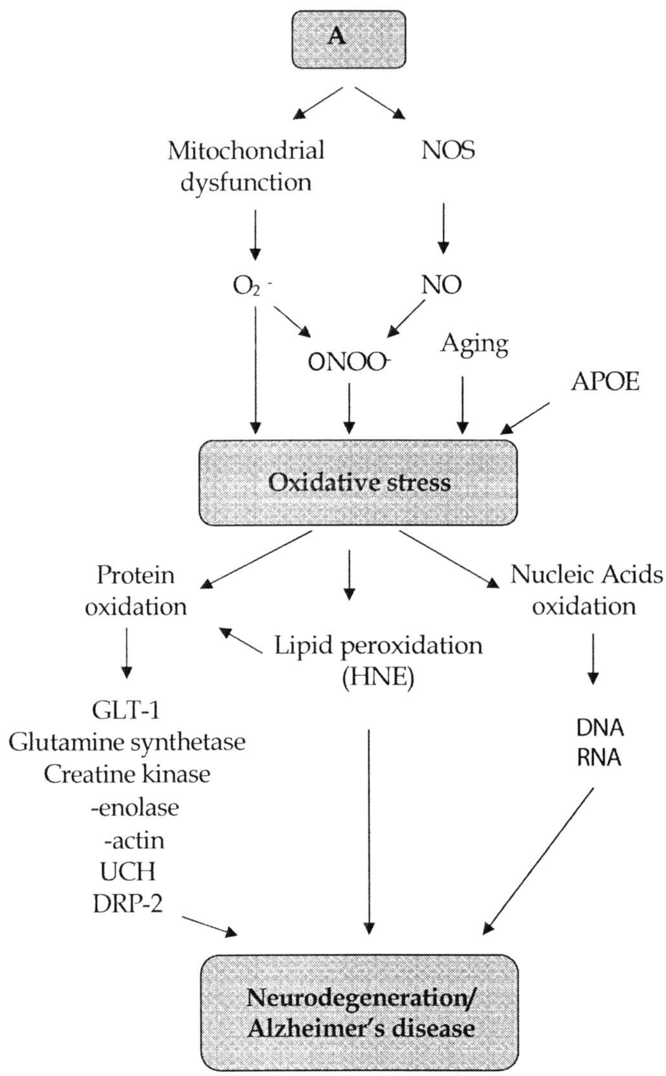

Figure 9. Schematic view of the role of oxidative stress in Alzheimer's disease death pathways. Brains from patients with Alzheimer's disease (AD) show relevant oxidative stress with reactive oxygen species (ROS) elevation, protein oxidation, lipid peroxidation and nucleic acid oxidation, apparently due to the action of amyloid β-peptide (Aβ). This peptide is capable of inducing nitric oxide synthase (NOS) as well, aggravating the production of ROS. 4-hydroxy-2-nonenal (HNE) formation, as

consequence of lipid peroxidation, has also been reported, as well as protein-bound HNE accumulation. Not only HNE but also ROS can damage proteins. Characteristic proteins damaged in AD include glial glutamate transporter (GLT-1), glutamine synthase, creatine kinase, α-enolase, β-actin, ubiquitin C-terminal hydrolase L-1 (UCH) and dihydropyrimidnase related protein-2 (DRP-2). Damage of each of these proteins may result in death (see text for details). The APOE allele is a risk for AD and it might be associated with HNE production. Aging is also considered a risk factor.

4.3. Stroke

Brain is particularly sensitive to deprivation of oxygen and glucose secondary to ischemia, due to its high metabolic rate and the complicated cell interaction. Actually, complete interruption of blood flow for only 5 minutes is sufficient to trigger neuronal death. Cerebral ischemia may be transient, followed by reperfusion, or essentially permanent, and it may affect part or the entire brain [18].

Ischemia induces mitochondrial dysfunction and consequent ATP depletion as well as ROS production in neurons [61]. ROS may originate single-strand DNA breakage, leading to the activation of repair enzyme poly (ADP-ribose) polymerase aggravating energy decline. Perturbations in cellular ATP levels and the increase in ROS production can easily originate an elevation of intracellular Ca^{2+}, which activates calpains and promotes neurotransmitters release such as Glu and dopamine. This further increases oxidative stress and may culminate in cell death [18].

Following ischemia, chelatable zinc ions translocate from nerve terminals to cell bodies, and this precedes neurodegeneration. A possible explanation for this is the potentiality of these metal ions to participate in Fenton reactions and increase oxidative stress. Ischemia and reperfusion also induce inflammatory responses that may exacerbate brain damage [95]. In microglia, cytokines and $NO^{\cdot}$ are produced, the latter by inducible NOS. In contrast, the NOS present in the endothelium may be protective in stroke, relaxing vascular smooth muscle cells and helping to preserve blood flow [18]. ROS are nevertheless capable of inhibiting this enzyme [19], as illustrated in Figure 10.

Neuronal death in ischemia occurs not only through necrosis, a fulminant cell death associated with swelling of cell body and organelles and plasma membrane failure, but also there is evidence of apoptosis in the infarcted brain. In fact, if it is true that failure of membrane transporters leads to Ca^{2+} accumulation, which is joined by sodium, chlorine and water entry and consequent swelling of dendrites and cell bodies, pointing to a possible death by necrosis, several studies have identified markers of apoptosis in stroke. These comprehend translocation of cytochrome *c* from the mitochondria to cytoplasm and activation of caspase-3 [18].

It should be emphasised that the only available treatment for stroke consists in the use of thrombolytic substances. However, the more the pathogenesis of this condition is cleared, the more attractive therapeutic targets seem to appear. Potential therapeutic targets in ischemia include calpains and ROS [96]. Zinc levels, also considered a threat in ischemia, may be lowered by improving its extrusion from nerve cells or providing intracellular sequestration. This could be achieved by dietary intake of chelators or by upregulating intracellular metallothioneins, zinc-binding proteins [97]. The final pathways leading to apoptosis are also likely to prove effective targets. For instance, ursodeoxycholic acid prevents mitochondrial perturbation and apoptotic death [98, 99]. Inflammation is a target for fighting stroke as well and COX-2 inhibitors have proven effective [100]. As far as excitotoxicity is concerned, several drugs have been developed in order to inhibit different Glu receptors [101]. It is also possible, in theory, to combine Glu receptors inhibitors and antiapoptotic substances [102], as well as thrombolytic substances with antioxidants or anti-excitotoxic drugs [103].

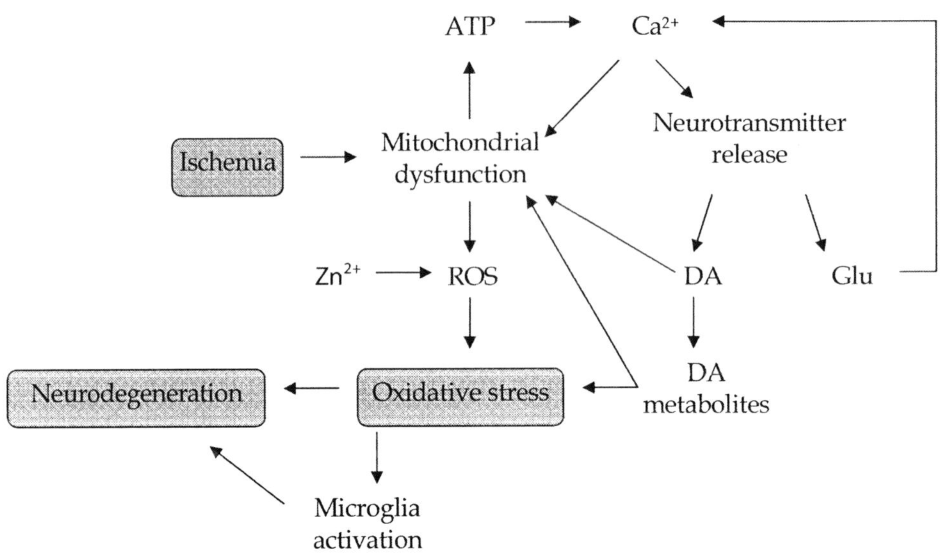

Figure 10. Schematic representation of oxidative stress role in stroke death pathways. Ischemia induces mitochondrial dysfunction and consequent adenosine triphosphate (ATP) depletion, as well as reactive oxygen species (ROS) production in neurons. Decreases in ATP levels can originate elevation of intracellular calcium (Ca^{2+}), which promotes neurotransmitters release such as glutamate (Glu) and dopamine (DA). This further increases oxidative stress and may culminate in cell death. Following ischemia, chelatable zinc ions (Zn^{2+}) translocate from nerve terminals to cell bodies, and this precedes neurodegeneration. A possible explanation for this is the potentiality of these metal ions to participate in Fenton reactions and increase oxidative stress. In microglia, cytokines and $NO^{\cdot}$ are produced, further contributing to neurodegeneration.

4.4. Prion Diseases

Prion diseases include the Bovine Spongiform Encephalopathy and its human variant Creutzfeldt-Jakob Disease. The most striking feature about prion diseases is the absence of a nucleic acid dependent transmission. Actually, the only necessary premise for the disease development is contact of the native prionic protein, designated PrP^c, with the abnormal form, usually named PrP^{Sc}. This results in a conformational change of the native protein, yielding more PrP^{Sc}, which is deposited as proteolysis resistant aggregates. There is no alteration, however in the amino acid sequencing [104].

Prion infected cells show increased ROS levels and decreased SOD, GPx and GR. This renders them particularly sensitive to oxidative stress and its mediated cell death [21]. Lipid peroxidation is also observed [20, 21]. There are several theories which may explain neurodegeneration in these diseases, all related to the conversion of PrP^c to PrP^{Sc}. To begin with, it is possible that this conversion may disturb copper metabolism, resulting in altered redox state and enzyme function [105]. It is also possible that the PrP^c itself has SOD activity and therefore, its conversion may result in a SOD decreased activity and a consequent ROS levels increase [106]. This is highly controversial and recent studies have proven otherwise [107, 108]. The conversion of PrP^c to PrP^{Sc} and affection of its functions may also modulate oxidative stress responses [109,110], through protein regulation or metal regulated transcription factors. This theory is supported by the fact that all GPX, GR and SOD, which share common pathways, are diminished in prion diseases. Finally, microglia activation may also contribute to the aggravation of oxidative stress [21], as seen in Figure 11.

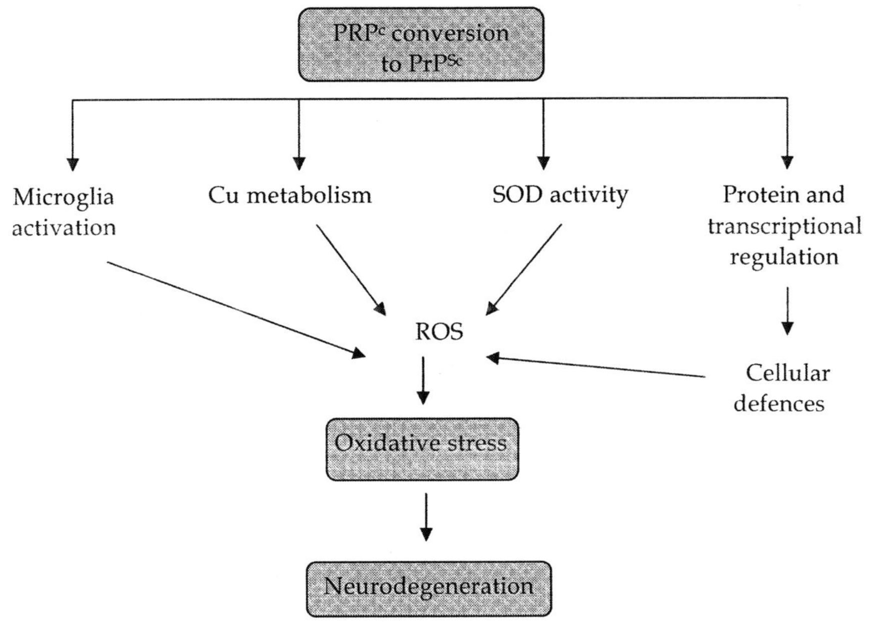

Figure 11. Representation of the role of oxidative stress in prion diseases death pathways. Conversion of the native prionic protein (PrPc) to the pathogenic prionic protein (PrPSc) may disturb copper (Cu) metabolism, resulting in altered redox state and enzyme function. If PrPc itself has superoxide dismutase (SOD) activity, its conversion may result in a SOD decreased activity and a consequent increase in reactive oxygen species (ROS) levels. The conversion of PrPc to PrPSc and affection of its functions may also modulate oxidative stress responses, through protein regulation or metal regulated transcription factors. This theory is supported by the fact that all GPX, GR and SOD are diminished in prion diseases. Finally, microglia activation may also contribute to the aggravation of oxidative damage.

The implication of oxidative stress in prion disease neurodegenerative processes is quite recent, but extremely relevant, once it opens new therapeutic perpectives. Approaches using copper chelants such as EDTA (ethylenediaminetetracetic acid) have been tested [111]. However, the most promising results were obtained with immunotherapy [112].

4.5. Bilirubin Encephalopathy

Unconjugated bilirubin (UCB) is a pigment resulting from heme catabolism. It is believed that UCB might be antioxidant [113] but beyond a certain threshold it is undoubtedly cytotoxic [114], particularly to the brain [115]. This toxicity may be reversible, when associated to physiologic neonatal hyperbilirubinemia, or irreversible when associated to Crigler Najjar type I syndrome [116], or severely jaundiced neonates, specially if premature. In the last case, UCB toxicity may result in permanent damage to the brain or even death (*kernicterus*). UCB is toxic to neurons [115] and apparently activates microglia, resulting in cytokine production [117]. This is also observed in astrocytes[118], even though the effect is not so marked.

UCB originates mitochondrial dysfunction [119] and oxidative stress in synaptosome vesicles, characterized by ROS elevation, decreased GSH/GSSG ratio, protein oxidation and lipid peroxidation [22]. Accumulation of Glu has also been reported, whether due to increased secretion [118] or decreased uptake [115,120]. It is conceivable that the decrease in ATP levels and consequent impairment of sodium/potassium ATPase [22] may, in part, be responsible for the increased levels of excitotoxic Glu. The neurotransmitter overstimulation of its receptors results in excessive Ca^{2+} intake by the nervous cell aggravating oxidative stress and promoting cell death (Figure 12), which may occur by necrosis or apoptosis.

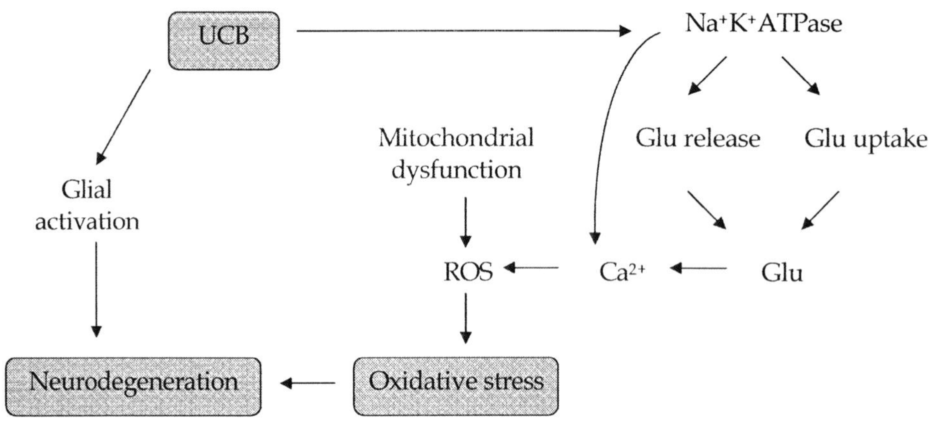

Figure 12. The role of oxidative stress in unconjugated bilirubin (UCB) death pathways. UCB originates mitochondrial dysfunction and reactive oxygen species (ROS) production in brain leading to oxidative stress. The impairment of sodium potassium (Na+, K+)-ATPase may be responsible for the increased levels of calcium (Ca2+) aggravating oxidative stress. Decline of the referred pump function may also contribute to the increase of excitotoxic glutamate (Glu), whether due to neuronal augmented secretion or decreased uptake. The neurotransmitter excessive stimulation of its receptors results in excessive Ca2+ intake by the nervous cell, aggravating oxidative stress and promoting cell death. UCB also activates microglia and astrocytes further contributing to neurodegeneration.

Dying cells exhibit phosphatidylserine at the outer leaflet of the plasma membrane [115]. This externalization may be a result of the reduced activity of aminophospholipid translocase [22], and probably constitutes a recognition signal of apoptotic cells for removal by phagocytosis.

Therapeutic approaches to treat or prevent UCB-induced oxidative damage in the brain hypothetically consist in the administration of antioxidants such as NAC or vitamin E. These are being investigated by us with growing interest because the population at greater risk of developing UCB encephalopathy, premature neonates, has increased in the last years, due to medical development and consequent improved survival.

5. Conclusion

Oxidative stress develops when ROS production, through mitochondrial respiration and oxygenase activity, outweighs the body's defence mechanisms, and results in the activation of complex pathways that culminate in cell damage and even death. The brain is particularly sensitive to oxidative damage due to its high metabolic rate and because differentiated neurons cannot divide and replace dead cells. This led the hypothesis that an association between oxidative stress and brain damage exists. In fact, oxidative stress has been surprisingly demonstrated to be a common link between several conditions such as PD, AD, stroke, prion diseases and UCB encephalopathy, where it is involved in neuronal injury. The brain is not a simple subject, nonetheless, but constitutes a complex environment where different cells, with unique configuration and function, interact with one another and respond differently to oxidative insult. This adds in difficulty to the study of the already complex neurodegenerative process. However, it is the study and comprehension of such complicated cell death pathways that points out specific therapeutic targets, allowing us to interfere with the disorder development, stopping it or, more optimistically, curing it. The application of such philosophy has resulted, for instance, in the usage of antioxidant drugs in animal models of neurodegenerative disorders, some of which successful enough to reach clinical trials. Results of the currently occurring clinical trials and new approaches originating in more investigation in this area will probably bring hope in the battle against such disabling diseases, improving patient's expectancy and quality of life.

References

[1] Halliwell, B; Gutteridge, JMR. *Free Radicals in Biology and Medicine.* 3rd Edition. New York: Oxford Science Publications; 1999.

[2] R. Dringen, Metabolism and functions of glutathione in brain. *Prog. Neurobiol.*, 2000, 62, 649-71.

[3] Dringen, R; Gutterer, JM; Hirrlinger J. Glutathione metabolism in brain. *J. Biochem.*, 2000, 267, 4914-6.

[4] Sies, H. Oxidative stress: oxidants and antioxidants. *Exp. Physiol.*, 1997, 82, 291-5.

[5] Yang, Y; Sharma, R; Sharma, A; Awasthi, S; Awasthi, YC. Lipid peroxidation and cell cycle signalling: 4-hydroxynonenal, a key molecule in stress mediated signalling, *Acta Biochim. Pol.*, 2003, 50, 319-36.

[6] Butterfield, DA; Castegna, A; Lauderback, CM; Drake, J. Evidence that amyloid beta-peptide-induced lipid peroxidation and its sequelae in Alzheimer's disease brain contribute to neuronal death. *Neurobiol. Aging*, 2002, 23, 605-64.

[7] Marnett, LJ; Riggins, JN; West, JD. Endogenous generation of reactive oxidants and electrophiles and their reactions with DNA and protein. *J. Clin. Invest.*, 2003, 111, 583-93.

[8] Kruman, I; Bruce-Keller, AJ; Bredesen, D; Waeg, G; Mattson, MP. Evidence that 4-hydroxynonenal mediates oxidative stress-induced neuronal apoptosis. *J. Neuroscience*, 1997, 17, 5089-100.

[9] Malecki, A; Garrido, R; Mattson, MP; Hennig, B; Toborek, M. 4-hydroxynonenal induces oxidative stress and death of cultured spinal cord neurons. *J. Neurochem.*, 2000, 74, 2278-87.

[10] Floyd RA. Antioxidants, Oxidative stress and degenerative disorders. *Proc. Soc. Exp. Biol. Med.*, 1999, 222, 239-45.

[11] Fahn, S; Sulzer, D. Neurodegeneration and neuroprotection in Parkinson disease. *NeuroRx*, 2004, 1, 139-54.

[12] Bové, J; Prou, D; Perier, C; Przedborski, S. Toxin-induced models of Parkinson's disease. *NeuroRx*, 2005, 2, 484-94.

[13] Rabacchi, SA; Friedman, WJ; Shelanski, ML; Troy, CM. Divergence of the apoptotic pathways induced by 4-hydroxynonenal and amyloid -protein. *Neurobiol. Aging*, 2004, 25, 1057-66.

[14] Perry, G; Cash, AD; Smith, MA. Alzheimer disease and oxidative stress. *J. Biomed. Biotechnol.*, 2002, 2, 120-3.

[15] Butterfield, DA; Lauderback, CM. Lipid peroxidation and protein oxidation in Alzheimer's disease brain: potential causes and consequences involving amyloid β-peptide-associated free radical oxidative stress. *Free Radic. Biol. Med.*, 2002, 32, 1050-60.

[16] Butterfield, DA. Amyloid β-peptide (1-42)-induced oxidative stress and neurotoxicity: implications for neurodegeneration in Alzheimer's disease brain. *Free Radic. Res.*, 2002, 36, 1307-13.

[17] Butterfield, DA; Boyd-Kimball, D; Castegna, A. Proteomics in Alzheimer's disease: insights into potential mechanisms of neurodegeneration. *J. Neurochem.*, 2003, 86, 1313-27.

[18] Lee, JM; Grabb, MC; Zipfel, GJ; Choi, DW. Brain tissue responses to ischemia. *J. Clin. Invest.*, 2000, 106, 723-31.

[19] Hoshino, S; Kikuchi, Y; Nakajima, M; Kimura, H; Tsuyama, S; Uemura, K; Yoshida, KY. Endothelial NO synthase (eNOS) phosphorylation regulates coronary diameter during ischemia-reperfusion in association with oxidative stress. *Free Radic. Res.*, 2005, 39, 481-89.

[20] Andreoletti, O; Levavasseur, E; Uro-Coste, E; Tabouret, G; Sarradin, P; Delisle, MB; Berthon, P; Salvayre, R; Schelcher, F; Negre-Salvayre, A. Astrocytes accumulate 4-hydroxynonenal adducts in murine Scrapie and humane Creutzfelt-Jacob disease. *Neurobiol. Dis.*, 2002, 11, 386-93.

[21] Milhavet, O; McMahon, HEM; Rachidi, W; Nishida, N; Katamine, S; Mangé, A; Arlotto, M; Casanova, D; Riondel, J; Favier, A; Lehmann, S. Prion infection impairs the cellular response to oxidative stress. *Proc. Natl. Acad. Sci. U.S.A.*, 2000, 97, 13937-42.

[22] Brito, MA; Brites, D; Butterfield, DA. A link between hyperbilirubinemia, oxidative stress and injury to neocortical synaptosomes. *Brain Res.*, 2004, 1026, 33-43.

[23] Rodrigues, CMP; Solá, S; Castro, RE; Laires, PA; Brites D. Perturbation of membrane dynamics in nerve cells as an event during bilirubin-induced apoptosis. *J. Lipid Res.*, 2002, 43, 885-94.

[24] Purves, D; Augustine, GJ; Fitzpatrick, D; Katz, LC; LaMantia, AS; McNamara, JO; Williams, SM. *Neuroscience.* 2nd Edition. Sunderland, U.S.A.: Sinauer Associates Inc.; 2001.

[25] Squire, LR; Bloom, FE; McConnell, SK; Roberts, JL; Spitzer, NC; Zigmond, MJ. *Fundamental Neuroscience.* 2nd Edition. San Diego, California: Academic Press; 2003.

[26] Butt, AM; Kiff, J; Hubbard, P; Berry, M. Synantocytes: new functions for novel NG2 expressing glia. *J. Neurocytol.*, 2002, 31, 551-65.

[27] Butt, AM; Hamilton, N; Hubbard, P; Pugh, M; Ibrahim, M. Synantocytes: the fifth element. *J. Anat.*, 2005, 207, 695-706.

[28] Siegel, GJ; Agranoff, BW; Fisher, SK; Albers, RW; Uhler, MD. *Basic Neurochemistry.* 6th Edition. U.S.A.: American Society for Neurochemistry: 1999.,

[29] Mizui T, Kinouchi, H; Chan, PH. Depletion of brain glutathione by buthionine sulfoximine enhances cerebral ischemic injury in rats. *Am. J. Physiol.*, 1992, 262, 313-7.

[30] Im, JY; Paik, SG; Han, PL. Cadmium-induced astroglial death proceeds via glutathione depletion. *J. Neurosci. Res.*, 2006, 83, 301-8.

[31] Bizzozero, OA; Ziegler, JL; De Jesus, G; Bolognani, F. Acute depletion of reduced glutathione causes extensive carbonylation of rat brain proteins. *J. Neurosci. Res.*, 2006, 83, 656-67.

[32] Zängerle, L; Cuenod, M; Winterhalter, KH; Do, JQ. Screening of thiol compounds: depolarization-induced release of glutathione and cystein from rat brain slices. *J. Neurochem.*, 1992, 59, 181-9.

[33] Guo, N; McIntosh, C; Shaw, C. Glutathione: new candidate neuropeptide in the central nervous system. *Neuroscience*, 1992, 51, 835-42.

[34] Lanius, RA; Shaw, CS; Wagey, R; Krieger, C. Characterization, distribuition, and protein kinase C-mediated regulation of [^{35}S]glutathione binding sites in mouse and human spinal cord. *J. Neurochem.*, 1994, 63, 155-60.

[35] Shaw, CA; Pasqualotto, BA; Curry, K. Glutathione-induced sodium currents in neocortex. *Neuroreport*, 1996, 7, 1149-52.

[36] Bukowska, B. Glutathione: its biosynthesis, induction agents and concentrations in selected diseases. *Med. Pr.*, 2004, 55, 201-9.

[37] Chen, Y; Swanson, RA. Astrocytes and brain injury, *J. Cereb. Blood Flow Metab.*, 2003, 23, 137-49.

[38] Walts, LT; Rathinam, ML; Schenker, S; Henderson, GI. Astrocytes protect neurons from ethanol-induced oxidative stress and apoptotic death. *J. Neurosci. Res.*, 2005, 80, 655-66.

[39] Dringen, R; Hamprecht, B; Broer, S. The peptide transporter Pep T2 mediates the uptake of the glutathione precursor CysGly in astroglia-rich primary cultures. *J. Neurochem.*, 1998, 71, 388-93.

[40] Kim, EJ; Park, YG; Baik, EJ; Jung, SJ; Won, R; Nahm, TS; Lee, BH. Dehydroascorbic acid prevents oxidative cell death through a glutathione pathway in primary astrocytes. *J. Neurosci. Res.*, 2005, 79, 670-9.

[41] Aschner, M; Sonnewald, U; Tan, KH. Astrocyte modulation of neurotoxic injury. *Brain Pathol.*, 2002, 12, 475-81.

[42] de Arriba, SG; Krugel, U; Regenthal, R; Vissiennon, Z; Verdaguer, E; Lewerenz, A; Garcia-Jorda, E; Palla, M; Camins, A; Munch, G; Nieber, K; Allgaier, C. Carbonyl stress and NMDA receptor activation contribute to methylglyoxal neurotoxicity. *Free Radic. Biol. Med.*, 2006, 40, 779-90.

[43] Anderson, CM; Swanson, RA. Astrocyte glutamate transport: review of properties, regulation, and physiological functions. *Glia*, 2000, 32, 1-14.

[44] Santos, MS; Moreno, AJ; Carvalho, AP. Relationship between ATP depletion, membrane potential, and the release of neurotransmitters in rat nerve terminals. An in vitro study under conditions that mimic anoxia, hypoglycemia, and ischemia, *Stroke*. 1996, 27, 941-50.

[45] Dringen, R; Hamprecht, B. Glutathione content as an indicator for the presence of metabolic pathways of amino acids in astroglial cultures. *J. Neurochem.*, 1996, 67, 1375-82.

[46] Daikhin, Y; Yudkoff, M. Glutamate and glutamine in the brain, compartmentation of brain glutamate metabolism in neurons and glia. *J. Nutr.*, 2000, 130, 1026-31.

[47] Schmuck, G; Roehrdanz, E; Haynes, RK; Kahl, R. Neurotoxic mode of action of artemisin. *Antimicrob. Agents Chemother.*, 2002, 46, 821-27.

[48] Schmuck, G; Roehrdanz, E; Tran-Thi, QH; Kahl, R; Schluter, G. Oxidative stress in rat neurons and astrocytes induced by Paraquat *in vitro*. *Neurotox. Res.*, 2002, 4, 1-13.

[49] Sponne, I; Fifre, A; Drouet, B; Klein, C; Koziel, V; Pinçon-Raymond, M; Olivier, JL; Chambaz, J; Pillot, T. Apoptotic neuronal cell death induced by the non-fibrillar amyloid-β peptide proceeds through an early reactive oxygen species-dependent cytoskeleton perturbation. *J. Biol. Chem.*, 2002, 278, 3437-45.

[50] Bolaños, JP; Heales, SJR; Land, JM; Clark, JB. Effect of peroxynitrite on the mitochondrial respiratory chain: differential susceptibility of neurones and astrocytes in primary culture. *J. Neurochem.*, 1995, 64, 1965-72.

[51] Mattson, MP. Excitotoxic and excitoprotective mechanisms – abundant targets for the prevention and treatment of neurodegenerative disorders. *Neuromolecular Med.*, 2003, 3, 65-94.

[52] Carafoli, E. Calcium signalling: a tale for all seasons. *Proc. Natl. Acad. Sci. U.S.A.*, 2002, 99, 1115-22.

[53] MacFarlane, M; Williams, AC. Apoptosis and disease: a life or death decision. *EMBO Rep.*, 2004, 5, 674-8.

[54] Harwood, SM; Yaqoob, MM; Allen, DA. Caspase and calpain function in cell death: bridging the gap between apoptosis and necrosis. *Ann. Clin. Biochem.*, 2005, 42, 415-31.

[55] Klein, JA; Ackerman, SL. Oxidative stress, cell cycle, and neurodegeneration. *J. Clin. Invest.*, 2003, 111, 785-93.

[56] Langley, B; Ratan, RR. Oxidative stress-induced death in the nervous system: cell cycle dependent or independent? *J. Neurosci. Res.,* 2004, 77, 621-9.

[57] Mattson, MP; Camandola, S. NF- B in neuronal plasticity and neurodegenerative disorders. *J. Clin. Invest.*, 2001, 107, 247-54.

[58] Mattson, MP; Meffert, MK. Roles for NF-kappa B in nerve cell survival, pasticity, and disease. *Cell Death Differ.*, 2006, 13, 852-60.

[59] Mattson, MP; Culmsee, C; Yu, Z; Camandola, S. Roles of nuclear factor kappa B in neuronal survival and plasticity. *J. Neurochem.*, 2000, 74, 443-56.

[60] Camandola, S; Poli, G; Mattson, MP. The lipid peroxidation product 4-hydroxy-2,3-nonenal inhibits constitutive and inducible activity of nuclear factor KB in neurons. *Brain Res. Mol. Brain Res.*, 2000, 85, 53-60.

[61] Almeida, A; Esteban, MD; Bolaños, JP; Medina, JM. Oxygen and glucose deprivation induces mitochondrial dysfunction and oxidative stress in neurones but not in astrocytes in primary culture. *J. Neurochem.*, 2002, 81, 207-17.

[62] Blanc, EM; Keller, JN; Fernandez, S; Mattson, MP. 4-Hydroxynonenal, a lipid peroxidation product, impairs glutamate transport in cortical astrocytes. *Glia*, 1998, 2, 149-60.

[63] Almeida, A; Almeida, J; Bolaños, JP; Moncada, S. Different responses of astrocytes and neurons to nitric oxide: the role of glycolytically generated ATP in astrocyte protection. *Proc. Natl. Acad. Sci. U.S.A.*, 2001, 98, 15294-9.

[64] Rahman, B; Kussmaul, L; Hamprecht, B; Dringen; R. Glycogen is mobilized during the disposal of peroxides by cultured astroglial cells from rat brain. *Neurosci. Lett.*, 2000, 290, 169-72.

[65] Buchet, K; Godinot, C. Functional F1-ATPase essential in maintaining growth and membrane potential of human mitochondrial DNA-depleted rho cells. *J. Biol. Chem.*, 1998, 273, 22983-9.

[66] Li, J; Baud, O; Vartanian, T; Volpe, JJ; Rosenberg, PA. Peroxinitrite generated by inducible nitric oxide synthase and NADPH oxidase mediates microglial toxicity to oligodendrocytes. *Proc. Natl. Acad. Sci. U.S.A.*, 2005, 102, 9936-41.

[67] Dewar, D; Underhill, SM; Goldberg, MP. Oligodendrocytes and ischemic brain injury. *J. Cereb. Blood Flow Metab.*, 2003, 23, 263-74.

[68] Ibarretxe, G; Sanchez-Gomez, MV; Campos-Esparza, MR; Alberdi, E; Matute, C. Differential oxidative stress in oligodendrocytes and neurons after excitotoxic insults and protection by natural polyphenols. *Glia*, 2006, 53, 201-11.

[69] Smith, KJ; Kapoor, R; Felts, PA. Demyelination: the role of reactive oxygen and nitrogen species. *Brain Pathol.*, 1999, 9, 69-92.

[70] Baud, O; Greene, AE; Li, J; Wang, H; Volpe, JJ; Rosenberg, PA. Glutathione peroxidase-catalase cooperavity is required for resistance to hydrogen peroxide by mature rat oligodendrocytes. *J. Neurosci.*, 2004, 24, 1531-40.

[71] Mronga, T; Stahnke, T; Goldbaum, O; Richter-Landsberg, C. Mitochondrial pathway is involved in hydrogen-peroxide-induced apoptotic cell death of oligodendrocytes. *Glia,* 2004, 46, 446-55.

[72] Vijitruth, R; Liu, M; Choi, DY; Nguyen, XV; Hunter, RL; Bing, G. cyclooxygenase-2 mediates microglial activation and secondary dopaminergic cell death in the mouse MPTP model of Parkinson's disease. *J. Neuroinflammation*, 2006, 3, 1-16.

[73] Kim, RH; Smith, PD; Aleyasin, H; Hayley, S; Mount, MP; Pownall, S; Wakeham, A; You-Ten, AJ; Kalia, SK; Horne, P; Westaway, D; Lozano, AM; Anisman, H; Park, DS; Mak, TW. Hypersensitivity of DJ-1-deficient mice to 1-methyl-4-phenyl-1,2,3,6-tetrahydropyrindine (MPTP) and oxidative stress. *Proc. Natl. Acad. Sci. U.S.A.*, 2005, 102, 5215-20.

[74] Fleming, SM; Fernagut, PO; Chesselet, MF. Genetic mouse models of parkinsonism: strengths and limitations. *NeuroRx*, 2005, 2, 495-503.

[75] Bower, JH; Maraganore, DM; MacDonnel, SK; Rocca, WA. Influence of strict, intermediate, and broad diagnostic criteria on the age- and sex- specific incidence of Parkinson's disease. *Mov. Disord.*, 2000, 15, 819-25.

[76] Tanner CM; Ottman, R; Goldman, SM; Ellenberg, J; Chan, P; Mayeux, R; Langston, JW. Parkinson disease in twins: an etiologic study. *J. A. M. A.*, 1999, 281, 341-6.

[77] Przedborski, S; Jackson-Lewis, V; Muthane, U; Jiang, H; Ferreira, M; Naini, AB; Fahn, S. Chronic levodopa administration alters cerebral mitochondrial respiratory chain activity. *Ann. Neurol.*, 1993, 34, 715-23.

[78] Whitehead, RE; Ferrer, JV; Javitch, JA; Justice, JB. Reaction of oxidized dopamine with endogenous cysteine residues in the human dopamine transporter. *J. Neurochem.*, 2001, 76, 1242-51.

[79] Zecca, L; Tampellini, D; Gatti, A; Crippa, R; Eisner, M; Sulzer, D; Ito, S; Fariello, R; Gallorini, M. The neuromelanin of human substantia nigra and its interaction with metals. *J. Neural Transm.*, 2002, 109, 663-72.

[80] Berg, D; Gerlach, M; Youdim, MB; Double, KL; Zecca, L; Riederer, P; Becker, G. Brain iron pathways and their relevance to Parkinson's disease. *J. Neurochem.*, 2002, 79, 225-36.

[81] Liu, Y; Peter, D; Roghani, A; Schuldiner, S; Prive, GG; Eisenberg, D; Brecha, N; Edwards, RH. A cDNA that supresses MPP^+ toxicity encodes a vesicular amine transporter. *Cell,* 1992, 70, 539-51.

[82] Wilms, H; Rosenberg, P; Sievers, J; Deuschel, G; Zecca, L; Lucius, R. Activation of microglia by human neuromelanin is NF-kappaB dependent and involves p38 mitogen-activated protein kinase: implications for Parkinson's disease. *FASEB J.*, 2003, 17, 500-2.

[83] Offen, D; Ziv, I; Sternin, H; Melamed, E; Hochman, A. Prevention of dopamine-induced cell death by thiol antioxidants: possible implications for treatment of Parkinson's disease. *Exp. Neurol.*, 1996, 141, 32-9.

[84] Mathews, RT; Yang, L; Browne, S; Baik, M; Beal, MF. Coenzyme Q_{10} administration increases brain mitochondrial concentrations and exerts neuroprotective effects. *Proc. Natl. Acad. Sci. U.S.A.*, 1998, 95, 8892-7.

[85] Shults, CW; Oakes, D; Kieburtz, K; Beal, MF; Haas, R; Plumb, S; Juncos, JL; Nutt, J; Shoulson, I; Carter, J; Kompoliti, K; Perlmutter, JS; Reich, S; Stern, M; Watts, RL; Kurlan, R; Molho, E; Harrison, M; Lew, M. Effects of coenzyme Q10 in early Parkinson disease: evidence of slowing of the functional decline. *Arch. Neurol.* 2002, 59, 1541-50.

[86] Cutler, RG; Kelly, J; Storie, K; Pedersen, WA; Tammara, A; Hatanpaa, K; Troncoso, JC; Mattson, MP. Involvement of oxidative stress-induced abnormalities in ceramide and cholesterol metabolism in brain againg and Alzheimer's disease. *Proc. Natl. Acad. Sci. U.S.A.*, 2004, 101, 2070-75.

[87] Castegna A, Aksenov, M; Thongboonkerd, V; Klein, JB; Pierce, WM; Booze, R; Markesbery, WR; Butterfield, DA. Proteomic identification of oxidatively modified proteins in Alzheimer's disease brain. Part I: creatine kinase BB, glutamine synthase, and ubiquitin carboxy-terminal hydrolase L-1. *Free Radic. Biol. Med.*, 2002, 33, 562-71.

[88] Boyd-Kimball, D; Castegna, A; Sultana, R; Poon, HF; Petroze, R; Lynn, BC; Klein, JB; Butterfield, DA. Proteomic identification of proteins oxidized by Aβ(1-42) in synaptosomes: implications for Alzheimer's disease. *Brain Res.*, 2005, 1044, 206-15.

[89] Gandy, S. The role of cerebral amyloid β accumulation in common forms of Alzheimer disease. *J. Clin. Invest.*, 2005, 115, 1121-29.

[90] Castegna, A; Aksenov, M; Thongboonkerd, V; Klein, JB; Pierce, WM; Booze, R; Markesbery, WR; Butterfield, DA. Proteomic identification of oxidatively modified proteins in Alzheimer's disease brain. Part II: dihydropyrimidinase-related protein 2, alpha enolase and heat shock cognate 71. *J. Neurochem.*, 2002, 82, 1524-32.

[91] Minghetti, L; Greco, A; Puopolo, M; Combrinck, M; Warden, D; Smith, AD. Peripheral reductive capacity is associated with cognitive performance and survival in Alzheimer's disease. *J. Neuroinflammation*, 2006, 3, 1-6.

[92] Fisher, AEO; Naughton, DP. Why nutraceuticals do not prevent or treat Alzheimer's disease. *Nutr. J.*, 2005, 4, 1-3.

[93] Duarte, TL; Lunec, J. When is an antioxidant not an antioxidant? A review of novel actions and reactions of vitamin C. *Free Radic. Res.*, 2005, 39, 671-86.

[94] Y. Christen, Oxidative stress and Alzheimer disease. *Am. J. Clin. Nutr.*, 2000, 71, 621-9.

[95] De Simoni, MG; Milia, P; Barba, M; De Luigi, A; Parnetti, L; Gallai, V. The inflammatory response in cerebral ischemia: focus on cytokines in stroke patients. *Clin. Exp. Hypertens.*, 2002, 24, 535-42.

[96] Markgraf, CG; Velayo, NL; Johnson, MP; McCarty, DR; Medhi, S; Koehl, JR; Chmielewski, PA; Linnik, MD. Six-hour window of opportunity for calpain inhibition in focal cerebral ischemia in rats. *Stroke*, 1998, 29, 152-8.

[97] Di Cello, F; Siddharthan, V; Paul-Satyaseela, M; Kim, KS. Divergent effects of zinc depletion in brain vs non-brain endothelial cells. *Biochem. Biophys, Res. Commun.*, 2005, 335, 373-6.

[98] Rodrigues, CMP ; Ma, X ; Linehan-Stieers, C ; Fan, G; Kren, BT; Steer, CJ. A novel role for ursodeoxycholic acid in inhibiting apoptosis by modulating mitochondrial membrane perturbation. *J. Clin. Invest.*, 1998, 101, 2790-99.

[99] Rodrigues, CMP; Ma, X; Linehan-Stieers, C; Fan, G; Kren, BT; Steer, CJ. Ursodeoxycholic acid prevents cytochrome *c* release in apoptosis by inhibiting mitochondrial membrane depolarization and channel formation. *Cell Death Dif.*, 1999, 6, 842-54.

[100] Candelario-Jalil, E; Mhadu, NH; González-Falcón, A; García-Cabrera, M; Muñoz, E; León, OS; Fiebich, BL. Effects of the cyclooxygenase-2 inhibitor nimesulide on cerebral infarction and neurological deficits induced by permanent middle cerebral artery occlusion in the rat. *J. Neuroinflammation*, 2005, 2, 1-11.

[101] Kew, JN; Trube, G; Kemp, JA. A novel mechanism of activity-dependent NMDA receptor antagonism describes the effect of ifenprodil in rat cultured cortical neurones. *J. Physiol.*, 1996, 497, 761-72.

[102] Ma, J; Endres, M; Moskowitz, MA. Synergistic effects of caspase inhibitors and MK-801 in brain injury after transient focal cerebral ischemia in mice. *Br. J. Pharmacol*, 1998, 124, 756-62.

[103] Seregi, A; Simmer, T; Schobert, A; Hertting, G. Characterization of cysteinyl-leukotriene formation in primary astroglial cell cultures. *J. Pharm. Pharmacol.*, 1990, 42, 191-3.

[104] Jackson, GS; Collinge, J. The molecular pathology of CJD: old and new variants. *J. Clin. Pathol. Mol. Pathol.*, 2001, 54, 393-99.

[105] Wong, BS; Chen, SG; Colucci, M; Xie, Z; Pan, T; Liu, T; Li, R; Gambetti, P; Sy, MS; Brown, DR. Aberrant metal binding by prion protein in human disease. *J. Neurochem.*, 2001, 78, 1400-8.

[106] Brown, DR; Wong, BS; Hafiz, F; Clive, C; Haswell, SJ; Jones, IM. Normal prion protein has an activity like that of superoxide dismutase. *J. Biochem*, 1999, 344, 1-5.

[107] Hutter, G; Heppner, FL; Aguzzi, A. No superoxide dismutase activity of cellular prion protein in vivo. *Biol. Chem.*, 2003, 384, 1279-85.

[108] Jones, S; Batchelor, M; Bhelt, D; Clarke, AR; Collinge, J; Jackson, GS. Recombinant prion protein does not possess SOD-1 activity. *J. Biochem.*, 2005, 392, 309-12.

[109] Rachidi, W; Vilette, D; Guiraud, P; Arlotto, M; Riondel, J; Laude, H; Lehmann, S; Favier, A. Expression of prion protein increases cellular copper binding and antioxidant enzyme activities but not copper delivery. *J. Biol. Chem.*, 2003, 278, 9064-72.

[110] Sakudo, A; Lee, DC; Li, S; Nakamura, T; Matsumoto, Y; Saeki, K; Itohara, S; Ikuta, K; Onodera, T. PrP cooperates with STI1 to regulate SOD activity in PrP-deficient neuronal cell line. *Biochem. Biophys. Res. Commun.*, 2005, 328, 14-9.

[111] Kim, NH; Choi, JK; Jeong, BH; Kim, JI; Kwon, MS; Carp, RI; Kim, YS. Effect of transition metals (Mn, Cu, Fe) and deoxycholic acid (DA) on the convertion of $PrP^{C}c$ to PrP^{res}. *FASEB J*, 2005, 19, 783-5.

[112] Korth, C; Peters, PJ. Emerging pharmacotherapies for Creutzfeldt-Jacob disease. *Arch. Neurol.*, 2006, 63, 497-501.

[113] Doré, S; Takahashi, M; Ferris, CD; Zakhary, R; Hester, LD; Guastella, D; Snyder, SH. Bilirubin, formed by activation of heme oxygenase-2, protects neurons against oxidative stress injury. *Neurobiology*, 1999, 96, 2445-50.

[114] Brito, MA; Silva, RFM; Brites, D. Bilirubin induces loss of membrane lipids and exposure of phosphatidylserine in human erythrocytes. *Cell Biol. Toxicol.*, 2002, 18, 181-92.

[115] Silva, RFM; Rodrigues, CMP; Brites, D. Rat cultured neuronal and glial cells respond differently to toxicity of unconjugated bilirubin. *Pediatr. Res.*, 2002, 51, 535-41.

[116] Jansen, PL. Diagnosis and management of Crigler-Najjar syndrome. *Eur. J. Pediatr.*, 1999, 158, 2, 89-94.

[117] Gordo, AC; Falcão, AS; Fernandes, A; Brito, MA; Silva, RFM; Brites, D. Unconjugated bilirubin activates and damages microglia. *J. Neurosci. Res.* 2006, [Epub ahead of print].

[118] Fernandes, A; Silva, RFM; Falcão, AS; Brito, MA; Brites, D. Cytokine production, glutamate release and cell death in rat cultured astrocytes treated with unconjugated bilirubin and LPS. *J. Neuroimmunol.*, 2004, 153, 64-75.

[119] Rodrigues, CMP; Solá, S; Silva, R; Brites, D. Bilirubin and amyloid-β peptide induce cytochrome c release through mitochondrial membrane permeabilization. *Mol. Med.*, 2000, 6, 936-46.

[120] Silva, R; Mata, LR; Gulbenkian, S; Brito, MA; Tiribelli, C; Brites, D. Inhibition of glutamate uptake by unconjugated bilirubin in cultured cortical rat astrocytes: role of concentration and pH. *Biochem. Biophys. Res. Commun.*, 1999, 265, 67-72.

In: Focus on Brain Research
Editor: Casey J. Resch, pp. 35-86
ISBN 1-600021-395-2

Chapter II

Neuroinflammation and Impairment of Glial Cell Regulation: Implications for the Pathogenesis of Alzheimer's Disease

Rommy von Bernhardi*
Department of Neurology, Faculty of Medicine,
Pontificia Universidad Católica de Chile.
Marcoleta 391, Santiago, Chile

Abstract

Inflammation is linked to neuronal cell death in several neurodegenerative diseases including Alzheimer's disease (AD). This association is based on neuropathological studies and the ability of anti-inflammatory compounds to modify the incidence of AD. Whether inflammation is a primary cause or only one of the consequences of the neurodegenerative process is still unclear. However, a growing body of evidence implicates inflammation in at least the progression of cellular injury that leads to neuropathology in AD. I propose that inflammation and cellular stresses associated with aging are primary initiating events in the development of AD. A dysregulated inflammatory response can elicit glial cell activation by compounds that are poorly reactive under normal conditions. Inflammation can also be the major cause of the defective handling of beta-amyloid (Aβ) and its precursor, amyloid precursor protein (APP), by glial cells, resulting in the accumulation and aggregation of Aβ and pathological pro-inflammatory activity. Glial cell activation in response to pro-inflammatory cytokines (which are up-regulated by many conditions including systemic diseases and local injury of the central nervous system) involves a complex range of

* Address correspondence to: Rommy von Bernhardi, MD, PhD. Neuroscience Laboratory, CIM. Department of Neurology, Faculty of Medicine, Pontificia Universidad Católica de Chile, Marcoleta 391, Santiago, CHILE, Tel 56-2-354 6936; Fax: 56-2-632 1924; E-mail: rvonb@med.puc.cl

responses that include morphologic, functional, and metabolic changes. Moreover there is an increased expression of genes involved in both nitric oxide (NO) and cytokine synthesis, agents that can cause neurodegeneration. We have found that microglial cell responses to APP and Aβ are low compared with those to pro-inflammatory cytokines and other mediators. However, activation is enhanced under pro-inflammatory conditions, indicating that glial cell responses to Aβ related proteins can be critically dependent on the priming of glial cells by pro-inflammatory factors. Here I review the evidence for inflammation's involvement in Aβ induced neurodegeneration. This evidence supports the theory that damage secondary to inflammation is not only a primary cause of neurodegeneration but also an initiator of the accumulation of Aβ in AD. I propose that the accumulation of Aβ, oxidative and nitrosative stresses, synaptic dysfunction and eventually neurodegeneration are related to the inflammatory status which itself is the primary event triggering dysregulation of glial cell activation. The responses of dysfunctional glia result in impaired neuronal function in AD, as well as in many slowly progressive neurodegenerative disorders.

I. Introduction: Alzheimer's Disease

AD is the most common dementia in old age. It is a degenerative disorder characterized by impaired mental function associated with neuronal dysfunction and selective neuronal loss (Selkoe, 2001). It is widely accepted that neurodegeneration is associated with the presence of increased amounts of β-amyloid peptide (Aβ) in the brain (Hardy and Selkoe, 2002). However, the generation of Aβ represents only one of the processes that determine the progression of AD (Figure 1).

The original report of AD lesions by Alois Alzheimer in 1906 was based on silver staining which revealed neurodegenerative changes. Now we know that the neuropathology of AD includes neuronal loss in the hippocampus and entorhinal and temporoparietal cortex (Goedert, 1998; Uylings and de Brabander, 2002). It is accompanied by the presence of two characteristic lesions, the deposition of fibrillar Aβ as insoluble extracellular aggregates forming the core of amyloid plaques - also named senile plaques (Arnold et al., 1998), and intracellular neurofibrillary tangles (Selkoe, 1996). Both aggregates are closely associated with abundant activated microglia and astrocytes (Sheng et al., 1995). Although senile plaques are mainly composed of fibrillar Aβ (Giulian et al., 1995; Sasaki et al, 1997), they are complex deposits also containing different proteins, metals and other reactive compounds (Dickson, 1997). However, do amyloid plaques and neurofribrillary tangles cause the disease, or are they themselves the result of earlier, pivotal events that lead to AD? It is unresolved if the accumulation of Aβ is the primary cause of AD or is a consequence of another pathophysiological change such as the pro-inflammatory environment associated with aging. The accumulation of Aβ does not necessarily constitute a senile plaque. Immune detection of Aβ reveals the presence of uncountable deposits in the brains of AD patients (and to a lesser extent in normal aged persons). These deposits are not fibrillar but rather diffuse plaques (Dickson, 1997; Duyckaerts et al., 1998) that are not associated with activated glia and/or inflammatory responses. Many postmortem studies show the presence of extensive Aβ deposits in the cerebral cortices of cognitively preserved elders (Dickson, 1997; Davis et al.,

1999); this suggests that the deposition of Aβ in the brain is a consequence of aging (Braak and Braak, 1997) and requires additional factors to become pathologic.

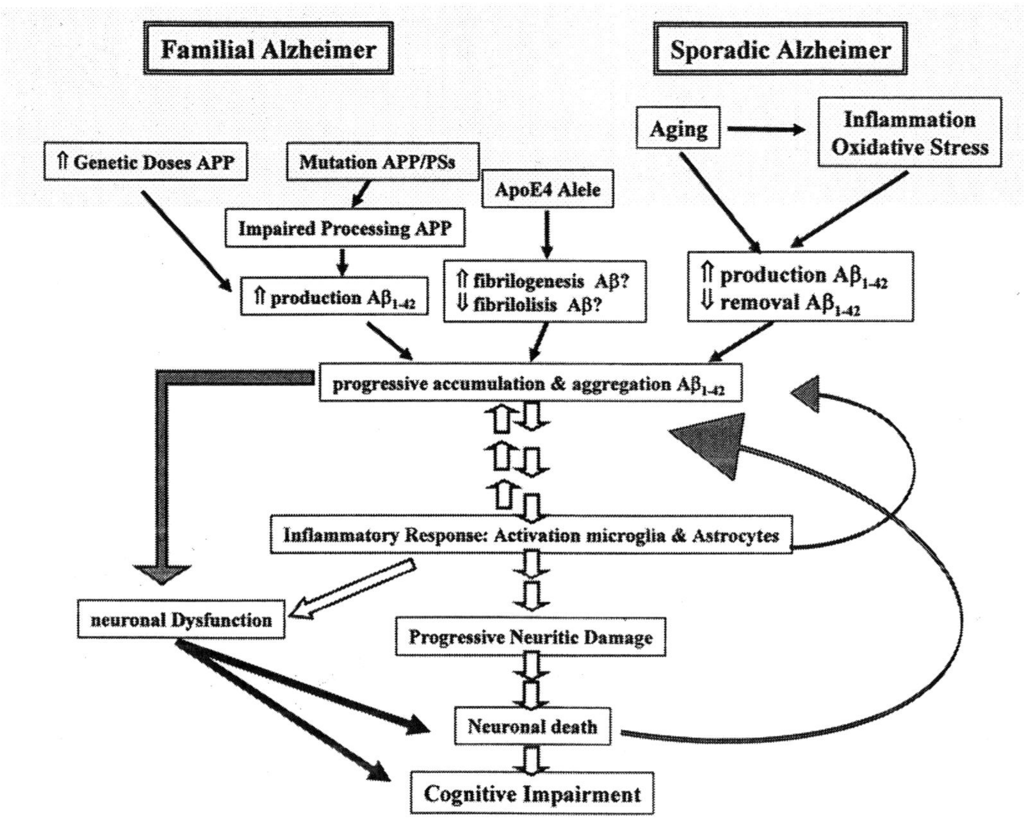

Figure 1. Neurobiological Mechanisms of Alzheimer's disease. Hypothetical sequence of events in the pathogenic cascade of AD. Aβ aggregates are present in familial and sporadic AD. However, whereas increased production of Aβ is consistently observed in the familial cases, in the sporadic form of the disease, it could depend on increased production or reduced removal of Aβ. Both mechanisms can be linked to aging, oxidative stress and the inflammatory status of the nervous tissue, which can also influence synaptic function and eventually neuronal viability.

In AD patients, the severity of dementia poorly correlates with the amount of amyloid plaques (Hardy, 1997). Although the presence of neurofibrillary tangles correlates better with cognitive impairment than the density of amyloid plaques (Braak et al., 1993; Jellinger, 1998), such tangles are a rather late event, with accumulation of Aβ preceding both the formation of neurofibrillary tangles and cognitive impairment (Naslund et al., 2000).

Aβ originates from the proteolytic processing of APP. APP is a type-I transmembrane protein with 3 isoforms that are ubiquitously expressed throughout the human body. One of these isoforms, APP-695 is especially abundant in nerve cells. Its function has not been clearly established, but it is associated with cell adhesion (Coulson et al., 2000), cell sorting (Kamal et al., 2001) and a transcription factor mediated by its cytoplasmic region (Ebinu and Yankner, 2002). Under nonpathological conditions, most APP is processed by a protease named α-secretase, thereby generating soluble APP (sAPP) which is proposed to have neurotrophic activity. In contrast, Aβ is produced by the sequential action of 2 proteases (Haass, 2004), β-secretase and γ-secretase (associated with presenilin activity); resulting in peptides of 39-43 amino acids (the 40 and 42 amino acid forms are the most abundant). Genetic studies have revealed mutations in 3 genes, APP and two presenilins (PS1 and PS2), which are associated with AD and the processing of APP. Because these mutations lead to increased production of Aβ, its overproduction is currently regarded as responsible for the development of AD (Hardy, 1997; Selkoe, 2000). These mutations associate with the familial forms of AD, accounting for 3-5% of AD patients. Over 95% of the patients suffering AD have the late-onset sporadic form. The presence of the Apo lipoprotein E4 (apoE4) allele is an important risk factor in developing sporadic AD (Roses, 1996; Farrer et al., 1997). Astrocytes synthesize most of the Apo E in the brain. ApoE is implicated in the clearance of Aβ, suggesting that an alternative explanation for the accumulation of Aβ in the brain is its decreased degradation. It is suggested that ApoE4 has less antioxidant activity than ApoE3 (Miyata and Smith, 1996). Clinical and experimental research has revealed that AD should not be considered as a unique nosological entity, as considerable heterogeneity exists in its risk factors, pathogenesis and neuropathological findings.

The way Aβ induces damage is still a matter of debate and several mechanisms are implicated. Aβ is toxic *in vitro* (Ramírez et al., 2005; Walsh et al., 1999) and *in vivo* (see Harkany et al., 2000). It induces cell damage to glial cells, both astrocytes and oligodendrocytes (Roth et al., 2005) and to neurons. Aβ neurotoxicity can directly act on neuronal cells (Geula et al., 1998; Lambert, 1998; Lorenzo and Yankner, 1994; Pike et al., 1991) or kill them indirectly through glial cell activation (Meda et al., 1995; Ishii et al., 2000). In both circumstances, astrocytes have a neuroprotective role. Activation of microglial cells by Aβ induces an inflammatory response and the release of potentially neurotoxic cytokines. Low concentrations of Aβ (2 μM), induce the expression of inducible NO synthase (iNOS) (von Bernhardi and Ramírez, 2001; von Bernhardi and Eugenín, 2004) and stimulate the release of NO by glial cells, similar to its upregulation by pro-inflammatory agents (Meda et al., 1995). Aβ is normally produced and secreted by various cells, including glial cells and neurons (Busciglio et al., 1993; Walsh et al., 2000) and is present in a soluble form in the human brain (Wisniewski et al., 1997). Glial reactivity increases under pro-inflammatory conditions, and it is useful to evaluate if such conditions cause the impairment in the ability of glial cells to handle and degrade APP or Aβ.

Aβ can have a role in synaptic function as well as in synaptic pathology (Kamenetz et al., 2003), since it accumulates in synapses, affects their function (Takahashi et al., 2004) and could induce synaptic degeneration (Gylys et al., 2004). In the early stages of AD the number and density of synapses decrease (Davies et al., 1987; DeKosky and Scheff, 1990; Bertoni-Freddari et al., 1996; Scheff and Price, 2003). There can also be functional alterations of hippocampal synapses prior to either neuronal degeneration or amyloid plaque formation (Selkoe, 2002; Yao et al., 2003).

APP transgenic mice show cellular, biochemical and electrophysiological evidence of synaptic dysfunction prior to the physical degeneration of synapses (Palop et al., 2003; Westphalen et al., 2003) and before Aβ deposition (Chapman et al., 1999). It has been proposed that this synaptic dysfunction is caused by soluble oligomeric assemblies of Aβ (Hartley et al., 1999; Kayed et al., 2003; Selkoe, 2001; Walsh et al., 2002). Addition of these soluble assemblies of Aβ to organotypic brain slices or neuronal cultures (Lambert et al., 1998; Wang et al., 2002) block long term potentiation (LTP). Also, microinjection of Aβ peptides into rat hippocampus impairs synaptic transmission (Stephan et al., 2001) and intraventricular microinjection into rats inhibits the maintenance of hippocampal LTP (Walsh et al., 2002).

Synaptic degeneration is a pathophysiological characteristic that AD shares with other neurodegenerative diseases (Cunningham et al., 2003; Li et al., 2003), suggesting the existence of common mechanisms. It has also been proposed that Aβ induces the release of excitatory neurotransmitters like glutamate, increasing intracellular calcium and the production of reactive oxygen species (ROS) in part because of mitochondrial injury (Canevari et al., 2004). The mechanism of action of memantine, a NMDA receptor blocker, recently approved for the treatment of AD (Parsons et al., 1999; Reisberg et al., 2003) is directed against this effect.

In contrast with a deleterious effect of Aβ on synapses, it has been suggested that Aβ is required for normal synaptic function and cell maintenance. Aβ also can have an essential function for neuronal survival because it has trophic and neuroprotective activities at physiological concentrations (Withson et al., 1989; Yankner et al., 1990) and at a nanomolar level blocks oxidative damage (Zou et al., 2002; Plant et al., 2003). This affords a protective role for Aβ (Lee et al., 2004) implying that Aβ accumulation in AD represents a protective mechanism triggered by the pathologic process (discussed in Koudinov and Berezov, 2004).

Pro-inflammatory cytokines are also associated with synaptic regulation and their participation is consistent with my proposition that glial inflammatory responses are responsible for AD pathology. The effects of cytokines on synapses have been considered insofar as their involvement in the pathogenic mechanisms of various CNS diseases. Tumor necrosis factor (TNF) is implicated in the pathogenesis of depression and also in the mechanism of action of antidepressant drugs (Reynolds et al., 2004). Different effects are often described depending on the concentration of cytokines or other environmental factors. Whereas interleukin 1β (IL-1β) under physiological conditions is involved in hippocampal-dependent memory and LTP, higher levels of IL-1β impairs memory and neural plasticity (Avital et al., 2003; Ross et al., 2003). IL-1β induces synaptic depression and neurotransmitter release, including acetylcholine and glutamate (Rada et al., 1991; Murray et al., 1997) and TNF-α can inhibit norepinephrine release in rat hippocampal slice experiments (Sperlagh et al., 2004; Reynolds et al., 2004). An intracerebroventricular injection of IL-1β attenuates LTP (Murray and Lynch, 1998a and 1998b). Changes in the cholinergic system, postulated in the cholinergic hypothesis as responsible for the cognitive deficit in AD patients (Bartus, 2000), also can be linked to changes in the inflammatory status of brain cells. Acetylcholinesterase (AChE) is able to activate glial cells, increasing NO production (von Bernhardi et al., 2003). There is also evidence that treatment with AChE inhibitors downregulates the expression of pro-inflammatory cytokines IL-6 and TNF-α in AD patients (Reale et al., 2004). In light of these observations, an alternative interpretation of the experiments involving oligomeric Aβ, is that the early synaptic dysfunction in AD is dependent on pro-inflammation, either secondary to aging or to the presence of preexisting pathology.

Immunotherapy in APP over-expressing mice, either as active (Schenk et al., 1999) or passive immunization (Bard et al., 2000), prevents the formation of amyloid plaques, neuritic dystrophy and glial reactivity (Schenk et al., 1999), and results in improved cognition (Younkin, 2001). Immunotherapy against Aβ in AD patients also results in reduced amyloid deposition and stabilized memory and cognitive declines (Hock et al., 2003; Nicoll et al., 2003; Gilman et al., 2004). Immunization improved microglial cell uptake and clearance of Aβ plaques. Although the cognitive improvement is interpreted as being associated with decreased levels of Aβ in the brain parenchyma (Nicoll et al., 2003), the improvement could also have resulted from the facilitation of Aβ uptake by microglial cells and the decrease of inflammatory activation of glial cells. As will be amplified below, the results are consistent with microglia having a key protective role, suggesting that, far from being detrimental, microglial activation can reduce AD pathology, as has been shown for the participation of microglia in the repair of injuries (Reichert and Rotshenker, 1996; Lazarov-Spiegler et al., 1998).

II. Aging and Alzheimer's Disease

The strongest risk factor for AD is aging. The prevalence of AD rises from 1-3% between ages 60 and 70, to 25-35% for people over 85 years old (Kukull and Bowen, 2002). Autopsy studies show an even higher incidence of neuropathological lesions and reveal that AD often occurs in conjunction with other pathologic lesions, especially vascular dementia and dementia with Lewy bodies (Fu et al., 2004). The overlap of pathologies suggests the existence of common pathophysiological mechanisms that probably have both genetic and environmental influences. Among such common mechanisms are glial cell activation and upregulation of inflammatory mediators. It is likely that various changes associated with aging cause pathology (Figure 2). When one of these changes occurs alone, its effect could be subtle but in combination, they can be highly detrimental (von Bernhardi, 2005). Brain hypo perfusion (de la Torre, 2002) and small brain infarctions (Vermeer et al., 2003) modify brain parenchyma homeostasis. A decreased blood flow can have various detrimental effects such as disruption of the production of energetic molecules and the maintenance of ionic gradients, increased intracellular calcium and increased production of oxygen free radicals. The microcirculation in AD patient brains has ultra structural and functional abnormalities which can further promote ischemia and hypoxia (Farkas et al., 2000). The susceptibility to ischemic brain damage is increased in APP transgenic mice (Zhang et al., 1997b). Cerebral ischemia induces a strong cerebral inflammatory response (Zhang et al., 1997a), APP expression (Kalaria et al., 1993) and increases amyloidogenic cleavage of APP (Wen et al., 2004; Yokota et al., 1996).

There are also general metabolic changes relevant for the likelihood of developing AD. Both the amount and distribution of cholesterol in plasma membranes changes with age, favoring the aggregation of Aβ (Wood et al., 2002). Animals exposed to a diet high in cholesterol have increased Aβ immune reactivity and plaques (Refolo et al., 2000). Cholesterol has been implicated in APP processing and generation of Aβ (Bodovitz et al., 1996; Simons et al., 1998), leading initially to disruption of synaptic plasticity and later to neurodegeneration (Michikawa, 2003). The prevalence of AD is reduced in individuals taking cholesterol-lowering drugs (Jick et al., 2000; Wolozin et al., 2000) and the risk of having AD is reduced by 40-70% (Wolozin et al., 2000; Rockwood et al., 2002).

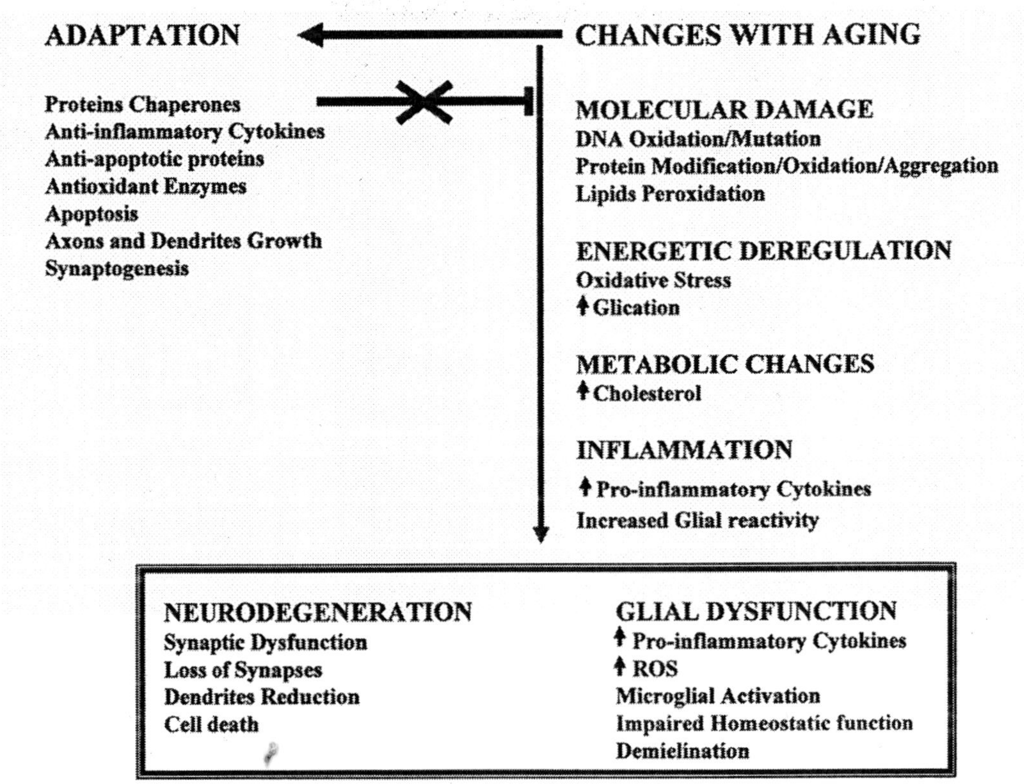

Figure 2. Cell Changes during Aging: Pathophysiological bases for Alzheimer's disease? Aging induces several functional and structural changes. These changes are often linked to inflammation and oxidative stress. In general, adaptation to aging allows preservation of an adequate function. However, if adaptation mechanisms fail or are insufficient to compensate aging changes, they will result in impairment of glial (further promoting activation and cell damage) and neuronal function.

However, the anti-inflammatory effect of some cholesterol lowering drugs like statins can also be important for their protective effects. Statins promote a shift in APP processing away from Aβ production (Fassbender et al., 2001), modify the expression of adhesion molecules, inhibit cytokine production, reduce astrocyte and microglial iNOS expression and have anti-oxidant effects both *in vivo* (Delanty et al., 2001; Vaughan and Delanty, 1999) and *in vitro* (Pahan et al., 1997).

An increased production of NO in the CNS (McCann, 1997) and increased oxidative stress have been associated with aging and to age-associated neurodegenerative diseases (Harmon, 1981). A shift in the balance either towards increased formation of free radicals formation or decreased antioxidant defenses results in increased oxidative stress that can lead to damage of lipids, DNA and proteins that induce cellular changes and neuronal degeneration. Lipid peroxidation is greater in AD individuals than that found in age-matched controls (Lovell et al., 2001; Markesbery et al., 1998). The nervous system is particularly sensitive to oxidative stress because of its metabolic dependence on aerobic carbohydrate metabolism, high oxygen consumption, relatively reduced levels of antioxidant defense mechanisms and the post mitotic nature of many brain cells. The redox state of a cell has an important influence on cell signaling. ROS function as second messengers in cytokine and growth factor-mediated cellular responses (Finkel, 2000; Lander, 1997). Age-related changes, such as the increased reactivity of microglial cells, suggest that cytokine secretion and oxygen radicals (Figure 3) can play a role in the development of age related neurodegenerative disorders including AD. Cortical plaques, mostly of the diffuse type, are abundant in a significant number of non-demented elderly individuals (Knopman et al., 2003). One of the characteristics of the diffuse plaques is that they do not show glial and inflammatory reactivity. Although it is not known how aging affects the response of microglia to activation-inducing stimuli such as Aβ, there is evidence that microglia from cognitively normal elderly individuals show several morphological abnormalities (Streit et al., 2004). Those microglial cells secrete several of the cytokines secreted by microglial cells exposed to Aβ (Lue et al., 2001a).

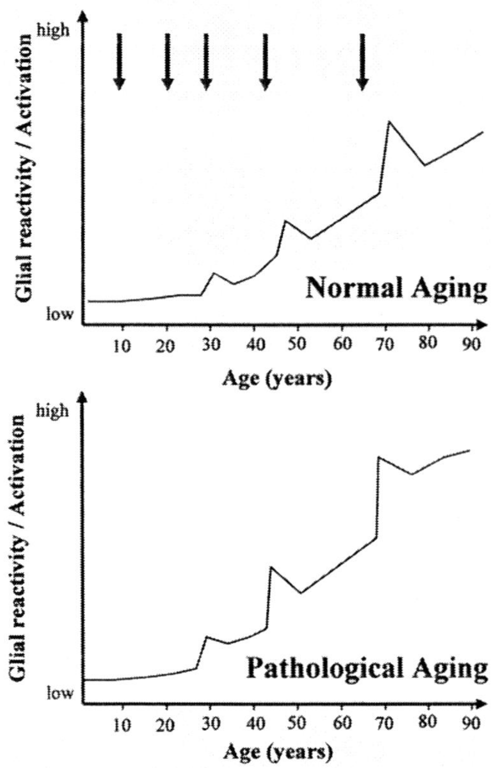

Figure 3. Changes in glial reactivity during Aging: Age-increase of glial reactivity may explain increased risk of developing a neurodegenerative disease observed in senile population. In aging, glial cells progressively increase their reactivity to various stimuli. Whereas a certain stimulus does not induce glial activation in young age, the same stimulus can induce a transient response and even can have a priming effect on glial cells in old age. In pathological aging a similar process occurs, however, remission of activation is not complete and glial cells persist partially activated after the stimuli, resulting in a permanent activation that progressively increases with each event of activation.

Production of growth factors declines during aging (Kelijman, 1991). Normal levels of insulin-like growth factor-1 (IGF-1) are required for the maintenance of several aspects of brain function, including the modulation of processes involved in energy supply to the cells, repair after injury, stimulation of neuron excitability, regulation of synaptic plasticity and some cognitive processes (Trejo et al., 2004). The reduction of growth factors is detrimental to aging and has multiple effects that contribute to frailty and morbidity. Especially relevant for neurodegenerative diseases are the increased cholesterol levels (Waters et al., 2003) and alterations in cognitive function (Aleman et al., 1999). Moreover, lack of IGF-1 is implicated in some of the pathophysiological changes characteristic to AD brains, including cholinergic dysfunction and neuronal amyloid toxicity. IGF-1 and growth hormone (GH) decreases are partially reversed by donepezil, one of the new cholinesterase-inhibitors (Obermayr et al., 2005), and could be related to its beneficial effects.

Synaptic function and the ability to generate LTP are attenuated in aged rats (Murray and Lynch, 1998a; 1998b). Hippocampal neurons of aged monkeys show alterations in morphology, intrinsic membrane proteins and signaling properties, defects that reduce their ability to perform various tasks regulated by the hippocampus that are similar to the tasks used to identify memory deficits in aged humans (Luebke and Rosene, 2003). Aging monkeys often exhibit cortical neuritic pathology, amyloid plaques and reactive glia within or at the periphery of senile plaques. In these animals, morphological abnormalities at synapses (both pre- and postsynaptic elements) precede the deposition of Aβ as well as activation of astrocytes and microglia. Neuronal perikarya/dendrites and reactive glia are primary sources of Aβ within amyloid plaques; nonfibrillar Aβ exists intracellularly prior to the appearance of extracellular deposits of Aβ (Martin et al., 1994). There are several changes in the response of brain cells exposed to Aβ associated with aging. Aβ does not induce damage when injected into young adult rhesus monkey's brain, whereas it causes neuronal loss and microglial reactivity in aged monkeys (Geula et al., 1998). This suggests that there are local changes in older monkeys that can facilitate the reactivity to Aβ. In contrast with cells from younger mice, adult mouse astrocytes exposed to Aβ do not secrete monocyte chemoattractant protein-1 (MCP-1), a chemokine important for the activation of astrocytes. Aging is also associated with changes of the ability of astrocytes to degrade Aβ and with differences in the oxidative potential of Aβ that can lead to a significant suppression of redox potential (Ham and Skoryna, 2004). In contrast with their neonatal counterparts, adult mouse astrocytes are efficient in clearing surface-bound Aβ and removing Aβ depositions in the brains of mice over expressing human APP (Wyss-Coray et al., 2003).

III. Inflammation and Alzheimer's Disease

Pro-inflammation refers to the presence of molecules or environmental conditions capable of activating glial cells. Glial activation results in diverse functional effects depending on its stage including proliferation and phagocytic transformation, and upregulation of several active molecules, release of cytokines and growth factors, and production of mediators such as NO (and eventually ROS).

There is strong evidence that inflammation is involved in multiple pathophysiological mechanisms (Lukiw and Bazan, 2000) and the progression of AD (Akiyama et al, 2000; Lim et al., 2000; McGeer and McGeer, 2001; Neuroinflammation Working group, 2000). Pro-inflammatory molecules and other soluble factors like NO are linked to cell damage and death (Liu et al., 2000). Clinicopathologic and neuroimage studies show that inflammation and microglial activation precede neuronal damage (Eikelenboom and van Gool., 2004). Oxidative stress and cell cycle abnormalities occur before the appearance of AD cytopathology (Zhu et al., 2004b). Cytokines such as IL-1β and Transforming growth factor-β (TGF-β) and the inducible form of cyclo-oxygenase (COX-2) are elevated in the brain of AD patients (Luterman et al., 2000; Ho et al., 2001). COX-2 rises rapidly in response to pro-inflammatory agents such as various interleukins and TNF and its expression is rapidly induced in glial cells and neurons in response to brain injury (Sairanen et al., 1998). The induction of COX-2 can result in the generation of ROS. Certain infections (Itzhaki et al., 2004) and neuroinflammation induced by lipopolysaccharide from gram negative bacteria (LPS) (Hauss-Wegrzyniak et al., 1998) increase intracellular accumulation of Aβ peptide (Sheng et al., 2003), inducing neuropathological changes similar to those found in AD brains. LPS and IFN-γ have been widely used as a model for neuroinflammation in various experimental approaches for the study of Alzheimer's and other neurodegenerative diseases (Hemmer et al., 2001; Herber et al., 2004; Sheng et al., 2003; Bal-Price and Brown, 2001). The exposure to LPS and IFN-γ is a strong pro-inflammatory signal for hippocampal cultures (Dell'Albani et al., 2001). It induces activation of microglial cells and astrocytes, and release of cytokines such as TNF (Bsibi et al., 2002; Bowman et al., 2003; Wesselingh et al., 1990). IFN-γ is induced in the nervous system as part of the inflammatory response (De Simone et al., 1998; Kiefer and Kreutzberger, 1990; Neumann et al., 1997; Olsson et al., 1994; Schmidt et al., 1990), and is associated with several degenerative diseases (Hofman et al., 1986; Traugott and Lebon, 1988).

McGeer and McGeer proposed that it is not the accumulation of Aβ but rather the inflammatory response to it that causes neuronal damage in AD. Several molecules with inflammatory and immune functions secreted by glial cells are associated with neurodegenerative disease lesions, revealing a state of chronic inflammation (McGeer and McGeer. 1995). Epidemiological evidence (Dziedzic et al., 2003; Sala et al., 2003) as well as experimental models of AD (Griffin and Mrak, 2002; Melton et al., 2003) show that pro-inflammatory conditions promote the development of AD. Nonsteroidal anti-inflammatory drugs decrease the risk of AD or delay its onset (Broe et al., 2000; Stewart et al., 1997), showing more benefits with long-term use (Etminan et al., 2003). The cellular bases for this neuroinflammation would be reactive microglia and astrocytes (Baltrons et al. 2002). Activated microglial cells synthesize and release short-lived cytotoxic factors including super oxide radicals and NO (Penkowa et al., 2000; Bal-Price and Brown, 2001). Oxidative stress induces apoptotic death of neurons (Vincent and Maiese, 1999b; Wittig et al., 2000). ROS, such as peroxynitrite and NO, can lead to cellular injury through inhibiting the electron transport chain of mitochondria. This blocks mitochondrial respiration (Yamamoto et al., 2002), peroxidation and leads to changes of cell membrane lipids (Salina et al., 2003), DNA damage (Vincent and Maiese, 1999a; Wang et al., 2003), and oxidation of proteins yielding protein carbonyl derivatives and nitrotyrosine. There is lipid peroxidation (Butterfield et al.,

2002), protein (Choi et al., 2003) and DNA (Lovell et al., 1999) oxidation, and nitrotyrosine in neuronal cell bodies (Allison et al., 2001) in the brains of AD patients. Colocalization of several oxidative stress markers with brain Aβ deposits is also found in APP transgenic mice (Smith et al., 1998). Moreover, there is evidence that ROS are also critical for pro-inflammatory gene expression, including iNOS, in glial cells (Pawate et al., 2004). Damage induced by NO can synergize with hydrogen peroxide increasing neuronal sensitivity to oxidative injury (de la Monte et al., 2003; Wang et al., 2003). The expression of iNOS is also increased in activated glial cells (Wallace et al., 1997; Dell'Albani et al., 2001). Various isoforms of iNOS are elevated in AD (Luth et al., 2002). The release of NO, both in culture (Bal-Price and Brown. 2001; Chao. 1996; Hu et al., 1997; Loihl and Murphy 1998; Wei et al., 2000) and *in vivo* (Loihl and Murphy 1998; Weldon et al., 1998) is one of the mechanisms by which activated glial cells exert their cytotoxicity. Other deleterious effects mediated by APP, include a decrease in antioxidant capacity (Takahashi et al., 2000) and activation of microglial cells (Barger and Harmon, 1997; Sondag and Combs, 2004).

The early stages of an inflammatory response can protect neurons (Wyss-Coray et al., 2002). However, failure to adequately deal with the inciting stimulus results in a chronic inflammatory response and the subsequent uncontrolled over-activation of microglia becomes detrimental, inducing the release of potentially toxic cytokines (Hanisch, 2002; Prinz et al., 1999; Schubert et al., 2000). Activated glial cells surrounding amyloid plaques or exposed to fAβ *in vitro* secrete pro-inflammatory molecules such as tumor necrosis factor-α, (TNF-α), IL-1β, MCP-1, RANTES, etc. (Hu and Van Eldik. 1999) and fatty acid metabolites such as eicosanoids (Liu and Hong, 2003), which participate in the early stages of neuroinflammation. Pro-inflammatory molecules secreted by microglia can be further linked to neurodegeneration because they increase the sensitivity of neurons to free radical exposure (Combs et al., 2001).

Besides their potential cytotoxic effects, pro-inflammatory cytokines such as TNF–α and IL-1β can also promote AD by stimulating the synthesis of APP (Goldgaber et al., 1989). A similar effect has been described for TGF-β1 on astrocytes (Rogers et al., 1999). Cytokines also stimulate the processing of APP towards the generation of Aβ (Blasko et al., 1999), including the stimulation of β-secretase activity (Sastre et al., 2003). Furthermore, high concentrations of cytokines promote the formation of fibrils and aggregation of Aβ. In contrast, the non-steroidal anti-inflammatory drug ibuprofen decreases cytokine-stimulated Aβ production (Blasko et al., 2001) and reduces neuritic plaques and inflammation in transgenic mice over expressing APP (Lim et al., 2000). Similarly, in APP-presenilin double transgenic mice, chronic treatment with a NO releasing anti-inflammatory drug, a derivative of flurbiprofen, reduces Aβ deposition. In all studies, long-term use of anti-inflammatory drugs reduce Aβ deposition, whereas the number of activated microglial cells increased in some studies (Jantzen et al., 2002) and decreased in others (van Groen and Kadish., 2005). Several non-steroidal anti-inflammatory drugs are able to decrease Aβ production (Weggen et al., 2001; Ericksen 2003), in some cases by changing secretase-γ-cleavage without affecting COX activity (its principal feature). These observations again suggest that pro-inflammatory conditions increase Aβ deposition. In apparent contradiction, intraperitoneal injection of APP transgenic mice with LPS, known to produce a sustained activation of microglial cells (Herber et al., 2005), reduces Aβ load during the first two weeks after the injection but Aβ returns to its basally high levels after an additional two weeks. There is no effect on aggregated Aβ plaques. These results indicate that activation of microglia results in the transient clearance of diffuse Aβ deposits. However, if microglial cells are unable to clear Aβ, it accumulates over time while microglial cells remain activated. (Herber et al., 2004).

IV. Microglial Cells: Their Participation in Neurodegenerative Diseases

Microglial cells are monocyte-derived immunocompetent cells that enter the CNS during late stages of embryonic development and become the innate immune system in the CNS (Giulian and Baker, 1986). They constitute the resident macrophage cells and orchestrate the inflammatory response to injury (Barron. 1995). Microglial cells synthesize several cytokines responsible for autocrine regulation and communication with neurons, astrocytes, and leukocyte infiltrates (Giulian and Corpuz, 1993; Hanisch, 2002). In unstimulated cultures, microglial cells express more anti-inflammatory cytokines such as TGF-β1 and IL-10 than the pro-inflammatory cytokines IL-1β, IL-6, IL-12, interferon gamma (IFN-γ) and TNF-α. When microglia are stimulated with a strong activator such as LPS or IFN-γ, pro-inflammatory cytokines are upregulated while anti-inflammatory cytokines are downregulated (Xiao et al., 1996).

There is an ongoing debate whether activation of microglia is beneficial or harmful (Giulian et al., 1993; Benveniste et al., 2001). Probably both are true since microglial activation is a gradual response. This gradual activation safeguards CNS homeostasis. Surveillance microglia sense and are attracted to the site of injury where they proliferate and undergo functional changes. These include the expression of *de novo* molecules that play a role in immunity and in the release of inflammatory modulators (Combs et al., 2001; Meda et al., 1999; Nguyen et al., 2002; Smits et al., 2000; 2002). Microglial cells are the early source of IL-1 following experimental CNS injury, infection or inflammation. The production of IL-1 by astrocytes usually occurs slightly later (Akiyama et al., 2000; Rothwell and Luheshi, 2000; Basu et al, 2004). Initially, activated microglia are neuro-protective against the pathological effects of noxious stimuli, (Streit, 2002). However, when activation is sustained, microglia-derived macrophages become cytotoxic (Eyüpoglu et al., 2003).

Autopsy studies show that microglial cells are activated in AD patients (Lue et al., 1996), (Rogers and Lue, 2001). Positron emission tomography reveals microglial activation in specific cortical regions in patients even in the early stages of AD (Cagnin et al., 2001), suggesting that microglial activation is an early event in the pathogenesis of the disease. The evolution of amyloid deposits towards mature amyloid plaques is associated with microglial cell activation (Sheng et al., 1997; Wegiel et al., 2000). The role of such microglial cells is controversial (Eikelenboom and van Gool, 2004), as there is evidence that they contribute to both the phagocytosis of Aβ (D'Andrea et al., 2004; Matsuoka et al., 2001), and to its deposition (Eikelenboom and van Gool, 2004; Wegiel et al., 2001).

It has been proposed that microglial cells associated with amyloid plaques generate oxidative stress. However, we have observed that generation of super oxide radicals in microglia exposed to Aβ *in vitro* is minor. This apparent contradiction could be resolved by considering that microglial cells *in vivo* are not activated solely by Aβ, but by a combination of pro-inflammatory stimuli. The ability of Aβ to elicit NO release is controversial (Forloni et al., 1997; Hu et al., 1998; Ishii et al., 2000; Meda et al., 1999). Some authors have shown that Aβ triggers NO production and pro-inflammatory molecules in microglial cells (Meda et al., 1995; 1996; Malchiodi-Albedi et al., 2001; Rogers et al., 2002) and astrocytes *in vitro* (Malchiodi-Albedi et al., 2001). We have observed that Aβ only induces a small increase in nitrite production in microglia, which is reduced in the presence of astrocytes. In contrast, high amounts of NO are produced by both astrocytes and microglial cells when activated by pro-inflammatory molecules (von Bernhardi et al., 2001), mediated through the activation of iNOS (Vegeto et al., 2001; Lieb et al., 2003). However Aβ together with LPS and IFN-γ have a synergistic effect on nitrite production, both in microglia and mixed glial cell cultures (von Bernhardi and Eugenín, 2004), as has been shown for other cell types (Baron et al., 2000; Goodwin et al., 1996). The reactivity of glial cells exposed to APP is low, but is potentiated in the presence of proinflammatory molecules (Figure 4).

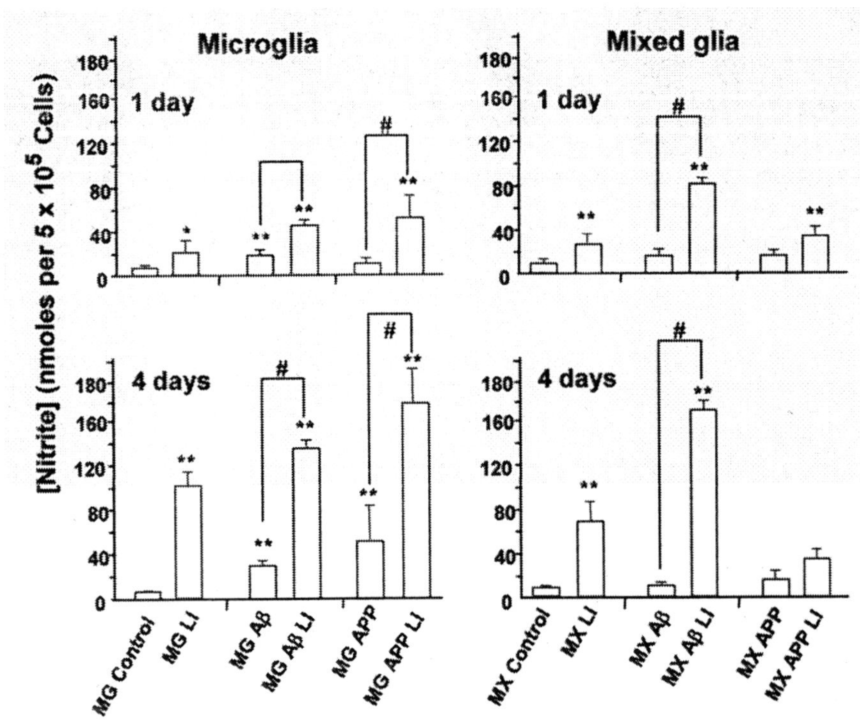

Figure 4. Glial reactivity to Aβ and APP - effect of pro-inflammatory molecules. Nitrite (NO_2^-, a stable product of NO) concentration in the supernatant of microglia (MG) or mixed glial cells (MX) after 1-4 days in culture was determined by the Griess assay. Non-stimulated cells (Control). Cells exposed to 2 μM fibrillar $A\beta_{1\text{-}42}$ (Aβ), 1 μg/ml LPS + 10 ng/ml IFNγ (LI), APP constructs (APP) or both (Aβ-LI), (APP-LI) for 1 and 4 days. APP corresponds to the segment of APP containing the 244 carboxyl-terminal amino acids. Bars correspond to the mean±SEM of the measurements in triplicate from 4-13 independent cell culture experiments for each condition. *, ** indicate conditions that were statistically different to the control condition. (*= $p<0.05$; **= $p<0.01$). # indicates conditions that were statistically different compared among indicated experimental conditions (Mann-Whitney U-Test).

The synergy between Aβ/APP and pro-inflammatory factors led me to postulate that microglia become reactive to Aβ/APP only under certain inflammatory conditions. These conditions can also determine whether microglia fail to respond to the modulatory effect of astrocytes and thus trigger an inflammatory response when exposed to Aβ. The implications for pathology are critical, since microglia and astrocytes are activated during aging and in most CNS diseases.

Upon activation by Aβ, independent of its aggregation according to some authors (Hashioka et al., 2005); microglial cells secrete pro-inflammatory cytokines (IL-1, IL-6 and TNF-α), chemokines (IL-8, MCP-1 and MIP-1) and the growth factor M-CSF (Lue et al., 2001b). In combination with pro-inflammatory cytokines such as TNF-α, NO is associated with neurodegeneration mediated by microglial cells (Hemmer et al., 2001). Microglial cells activated by pro-inflammatory molecules, in contrast with astrocytes, also produce large amounts of $O_2^{\cdot -}$ by the "respiratory burst" (Chao et al., 1995; Eikelenboom et al., 2002) which is associated with apoptotic cell death. Aβ (Streit, 2002; Hanisch, 2002) also activates and induces the respiratory burst of microglial cells (Akiyama et al., 2000). The generation of ROS by microglia is responsible for this cell's cytotoxic activity. ROS scavenger agents can prevent cellular injury mediated by activated microglia (Tanaka et al., 1994). However, microglial mediated neurotoxicity in AD does not solely depend on the generation of ROS, but also occurs through the production of cytokines and their effects on neighboring cells, including neurons (Benzing et al., 1999; Mehlhorn et al., 2000). Microglial cells mediate hippocampal cell death induced by the combination of LPS and IFN-γ, but only after long exposures of several days (Saud et al., 2005), suggesting that toxicity depends on the *de novo* synthesis of molecules by hippocampal cells. Microglial cells exposed to Aβ also secrete soluble factors capable of inducing cell death. The cytoxicity of the conditioned media is preserved for several days, suggesting that it does not depend on short-lived compounds such as radicals (von Bernhardi and Eugenín, 2004).

V. Astrocytes: Their Participation in Neurodegenerative Diseases

Astrocytes are the most abundant cells of the brain. They are the structural and trophic support of the CNS (Kirchhoff et al., 2001) as well as multi-functional cells needed for neuronal housekeeping, synapse formation and modulation (Nedergaard, 1994). Astrocytes play an important role in inflammatory processes (Wyss-Coray et al., 2003) and their activation in many diseases is associated with the induction of chemokines and inflammatory cytokines (Johnstone et al., 1999) similar to those described for microglial cells. Besides cytokines, astrocytes also secrete neurotrophic factors such as nerve growth factor (NGF), S100β, brain derived growth factor (BDNF) and neurotrophins (Aguado et al., 1998; Blondel et al., 2000; Hock et al., 2000; Mrak and Griffin, 2001). However, astrocytes are also damaged, sometimes preceding neuronal damage, as in ischemic brain injury (Liu et al., 1999).

In AD brains, astrocyte activation is prominent around amyloid plaques and diffuse Aβ deposits. They work cooperatively with microglia and contribute to the local inflammatory response in senile plaques (Vincent et al., 1998). Astrocytes and microglia react in a concerted manner exerting mutual regulatory activity over each other (Giulian et al., 1994; Ramírez et al., 2005; Saud et al., 2005). Astrocyte activation induces production of growth factors that promote growth and activation of microglial cells (Lee et al., 1994). Astrocytes have recently become a focal point in research on neurodegenerative diseases because of their neuroprotective role following a variety of cell injuries such as excitotoxicity, oxidative stress, ischemia and Aβ toxicity.

In AD experimental models, astrocytes modulate microglial cytotoxicity (von Bernhardi and Ramírez, 2001; von Bernhardi and Eugenín, 2004; Ramírez et al., 2005). Evidence is still contradictory as to whether reactive astrocytes accumulating at sites of Aβ deposition have damaging or neuroprotective functions (Wyss-Coray and Mucke. 2002). Whereas astrocytes do not produce ROS (Husemann and Silverstein. 2001; Husemann et al., 2001) and have neuroprotective properties (Smits et al., 2001; Sortino et al., 2004; von Bernhardi and Eugenín, 2004; Wyss-Coray et al., 2003), they secrete pro-inflammatory mediators such as IL-1, MCP-1, RANTES and TNF-α (Smits et al., 2001), produce NO when stimulated by Aβ, and show deleterious effects (Abramov et al., 2003; Malchiodi-Albedi et al., 2001; Nagele et al., 2004) in AD. On the other hand, astrocytes isolated post mortem from AD patients produce high levels of anti-inflammatory cytokines such as TGF-β1 (Blasko et al., 2004) and reactive astrocytes from AD-transgenic mice secrete TGF-β1 and IL-10 (Apelt and Schiebs, 2001). Because TGF-β1 production can be inhibited by TNF-α (Blasko et al., 2004), it is possible that pro-inflammatory factors downregulate and neutralize astrocytes' protective anti-inflammatory responses. In animals, the effects of these opposing actions of astrocytes on neurons probably depend on the presence of other glial cells, environmental conditions, and the type of neuron.

VI. Modulation of Glial Activation: Cross Talk among Glial Cells and Neurons

Glial cells and neurons influence each other under physiological and pathological conditions (Figure 5). Neurons have an inhibitory role on glial activation *in vitro*, by regulating the production of soluble factors and many proteins associated with reactive gliosis (McMillian et al., 1994). Both membrane-bound and secreted factors are provided by neurons, including OX2, neurotrophins and cytokines (Hoek et al., 2000). Active healthy neurons provide inhibitory factors suppressing the inflammatory responses of microglia and astrocytes (Neumann and Wekerle, 1998), but damaged neurons can induce glial activation (Sudo et al., 1998). Hippocampal cultures secrete soluble factors, such as TGF-β capable of modulating $O_2^{\cdot -}$ and NO production by glial cells. However, neurons also produce inflammatory mediators, including eicosanoids (Pasinetti, 1998), C-reactive protein, amyloid protein, and complement factors (Yasojima et al., 2000); their release, further stimulates a glial inflammatory response. As with most homeostasis response systems, neurons provide feedback loops maintaining the magnitude and duration of glial activation and restoring glial resting conditions once these functions are completed.

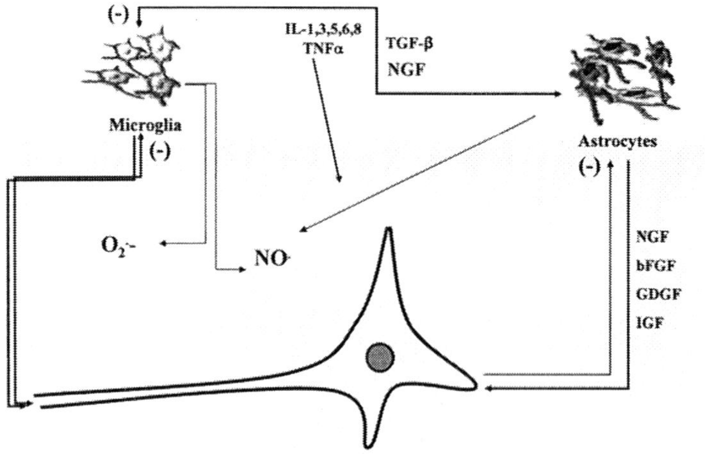

Figure 5. Modulation in the CNS: Neuron-glia Interaction. The homeostasis of the nervous tissue is maintained by a finely tuned cross talk of glial cells and neurons. The same cells are responsible for orchestrating the reaction to injury. Astrocytes (and also microglia) secrete growth factors and cytokines. There is a reciprocal structural and functional regulation of glial cells and neurons by the factors secreted by these cells. Pro-inflammatory cytokines are deemed as potentially responsible for cell damage in neurodegenerative diseases and induce the production of NO and ROS, which have been associated to cell damage. Factors, such as TGF- , induced by the pro-inflammatory conditions, modulate the activation of glial cells. Astrocytes and microglial cells modulate their activity and neurons have an inhibitory regulatory role on glial activation. Neurodegeneration could depend on the inability of microglial cells to respond to the modulatory effect mediated by neurons and astrocytes. As an alternative, the modulatory mechanism itself could fail to be produced.

Conversely, both microglia and astrocytes produce several trophic factors and extracellular matrix molecules that are needed by neurons for their maintenance and function. Such interactions are also meaningful in AD (Griffin et al., 1998). Thus, glial-neuronal interactions and the modulation of the cytokines they secrete could have a role in the progression of AD. Injuries or conditions affecting neuronal function can impair their inhibitory effect on a glial inflammatory response, which in turn would further impair the nervous system.

Glial activation, a CNS injury hallmark, can both impair and promote neuronal survival. Glial cells release (see above) a variety of molecules, such as ROS, nitrogen intermediaries, pro-inflammatory mediators like cytokines and chemokines (Meda et al., 2001), complement factors, etc., that can induce cell damage (Akiyama et al., 2000; Eikelenboom et al., 2002); they also secrete molecules that promote cell survival in culture. Apoptotic cell death is observed when these factors are inhibited, removed or their actions blocked (Rubin, 1998). Microglial cells secrete inflammatory cytokines needed to activate astrocytes in order for them to secrete growth factors able to promote the survival and recovery of damaged neurons. The complex interaction between astrocytes and microglia and their reciprocal modulatory effects makes it very difficult to distinguish their specific activities. This duality of protective and damaging effects suggests that astrocytes can play a neuro-protective role early in activation. However, in later stages, impairment of their modulatory capability or their inability to modulate a persistent activation could lead to neurocytotoxicity.

Microglial cytotoxic activity is modulated in various ways, including the induction of antioxidant enzymes like catalase or super oxide dismutase and anti-inflammatory cytokines such as IL-1Ra, IL-4, IL-10 and TGF-β1. Their activity also is modulated by astrocytes. Evidence supports a role for astrocytes in removing toxic factors secreted by microglial cells and Aβ clearance (Shaffer et al., 1995; Wyss-Coray et al., 2003). Astrocytes affect the interaction of microglia with Aβ, through reducing their microglial activation (von Bernhardi and Ramírez, 2001) and altering the phagocytic capability of microglial cells (DeWitt et al., 1998). Astrocytes also attenuate the production of oxygen radicals and TNF-α by monocyte-derived macrophages (Smits et al., 2001). Astrocytes, as well as soluble factors secreted by astrocytes activated with pro-inflammatory molecules, abolish Aβ cytotoxicity, both directly (Ramírez et al., 2005) and indirectly through microglial activation (von Bernhardi and Eugenín, 2004). TGF-β1 can be one of the modulating factors, since modulation is abolished when conditioned media from glial cells is treated with an antibody specific for active TGF-β1 (Herrera-Molina and von Bernhardi. 2005). This suggests that the physical presence of astrocytes is not required and that inhibition of cytotoxicity is probably due to soluble factors secreted by them.

Diverse functions are differentially affected during activation of microglia and astrocytes. This is consistent with Aβ and the combination of LPS and IFN-γ activating glial cells through different mechanisms, as observed for other cytokines (Forloni et al., 1997; Vincent et al., 2002). Several pathways seem to be involved both in microglial (Bales et al., 1998) and astrocyte activation by Aβ (Akama et al., 1998), including tyrosine kinase-dependent pathways (Combs et al., 2001; McDonalds et al., 1998) and NF-kappaB. Several of the pathways activated by Aβ are common with those activated by IFN-γ and LPS (Nguyen et al., 2002). However, only the combination of LPS and IFN-γ is able to induce complete activation of macrophages through the JAK-STAT pathway (Kovarik et al., 1998). Taken together, these results suggest that astrocytes are pivotal in the modulation of inflammation of the CNS and have major regulatory effects on microglia and neurons exposed to Aβ. Impairment of the regulatory functions performed by activated astrocytes could represent an important pathogenic mechanism for neurodegenerative diseases.

VII. Pro- and Anti-Inflammatory Cytokines

Cytokines are constitutively expressed in glial and neuronal cells and exert actions on them in the normal brain in the absence of pathophysiological stimuli, presumably participating in complex behaviors such as sleep and feeding and modulating various functions, including cell homeostasis, metabolism, synaptic function and plasticity, neural transmittion. When upregulated in response to injury, infection, or various neurodegenerative and inflammatory conditions, primarily astrocytes and microglial cells, but also neurons, produce pro-inflammatory mediators. These include chemokines, cytokines such as TNF-α, IL-1, IL-6, prostaglandins, IFN-γ (De Simone et al., 1998), and anti-inflammatory molecules, among which TGFβ is one of the most important.

Pro-inflammatory cytokines not only amplify the inflammatory response but are also protective (Benveniste, 1998). Glial cells synthesize and at the same time are targets for many cytokines, and release neuroactive factors in response to cytokine stimulation. Thus, the end result of the response to injury depends on both the profile of the secreted cytokines and other factors present in the immediate cellular environment. IL-1β modulates inflammation and multiple cell functions in basal conditions. Its level under nonstimulated conditions is low and restricted to certain areas of the CNS such as the hypothalamus, hippocampus and cerebral cortex. It is expressed by neurons and glial cells. It increases during aging, associated with the increased reactivity of glial cells. IL-1β is associated with multiple CNS injuries. Although considered a pro-inflammatory cytokine, IL-1β can either contribute to or limit neuronal damage or death (Rothwell and Luheshi, 2000). IL-1 can exert beneficial effects, particularly when released in modest amounts (Mrak et al., 1995; Basu et al., 2004). IL-1β promotes remyelinization (Arnett et al., 2003) and induces the upregulation of growth factors needed for the survival of various neuronal populations (De Kosky et al., 1996; Herx et al., 2000). Administration of IL-1β increases TGF-β1 in the brain and has a synergic effect with the neuroprotective activity of NGF (Chalazonitis et al., 1992; Vincent et al., 1997). In AD, IL-1β expression is induced early during plaque formation (Akiyama et al., 2000; Rothwell and Luheshi, 2000). The negative actions and beneficial effects exerted by IL-1β is a controversial area. It has been proposed that IL-1β induces or potentiates cytotoxicity because both IL-1β and TNF-α induce iNOS expression and NO production through the activation of extracellular signal-regulated kinase (ERK) and NFkB (Marcus et al., 2003). Activation of the IL-1 receptor IL-1RI leads to the translocation of NFkB to the nucleus and the activation of the MAPK pathways ERK, p38 MAPK and JNK1 (Rothwell and Luheshi, 2000). The expression of IL-1RI is also regulated by cytokines (Friedman, 2001).

TNF-α is expressed at the mRNA and protein levels in several regions of the CNS in pathology, but its expression in normal healthy brain is controversial (Vitkovic et al., 2000). It is proposed that IL-1β and TNF-α signals (both activate NFkB and JNK signal transduction pathways) are integrated within the cell and interact in regulating several cellular functions such as sleep and integration of autonomic responses. TNF-α also can modulate neurons, having effects on synaptic activity and release of neurotransmitters. Depending on the physiological state of the target, TNF-α exerts pleiotropic effects on neurons, including toxic (Talley et al., 1995), protective (Cheng et al., 1994; Barger et al., 1995) and as a modulator of neural transmission (Barone et al., 1997). When upregulated by pro-inflammatory stimuli or injury, TNF-α is involved in cell death, a function that would be a key factor to understand its effects on neurons. TNF-α has a key role as mediator of tissue injury and inflammation, being implicated in the pathogenesis of several neurological diseases. However, in animal models for demyelization diseases and traumatic brain injury it also has a protective role (Arnett et al., 2001), apparently mediated by its effect on various cellular receptors. TNF-α could be pro-inflammatory at early acute stages, but anti-inflammatory during the chronic phase. IFN-γ is considered as the central mediator of autoimmune diseases, but there is evidence suggesting it has a protective effect in animal models of autoimmune disease (Willenborg et al., 1996). IFN-γ prevents cell death of neurons deprived of NGF (Chang et al., 1990). In addition, the combination of LPS and IFN-γ prevents Aβ-induced apoptosis in hippocampal cell cultures. The beneficial effect of LPS and IFN-γ is associated with the increased expression and secretion of TGF-β1 by hippocampal cultures and depends on the activation of astrocytes, inasmuch as its protective effect is absent in hippocampal cultures depleted of astrocytes (Ramírez et al., 2005).

TGF-β1 has various functions (Mouri et al. 2002; Hayashida et al., 2004; Ishida et al. 2004), including prominent roles in tissue development, homeostasis and repair (Akiyama et al., 2000). In the central nervous system TGF-β1, as well as other endogenous neurotrophic factors, increase with age (Bye et al., 2001; Harry et al., 2000), apparently in response to changes in glial activation during aging (Nichols. 1999), whereas TGF-β1 decreases in several non neuronal tissues (Matsunaga et al, 2003). TGF-β1 is a good candidate for a protective cytokine. It is synthesized in the brain in response to various insults such as ischemia (Krupinski et al., 1996), is involved in various neurodegenerative pathologies and has been implicated as one of the factors secreted by astrocytes mediating neuroprotection (Dhandapani et al., 2003). Also, treatment with exogenous TGF-β1 has a protective effect on Aβ–induced neurotoxicity (Prehn et al., 1996). There are low concentrations of TGF-β1 in normal brain tissue, whereas its expression is increased in activated glial cells in the injured brain and in several pathologic states (O'Brien et al., 1994; Zhu et al., 2000, Meda et al., 2001; Chen et al., 2002). Hippocampal cells and activated astrocytes secrete TGF-β1 both *in vivo* and *in vitro* (Flanders et al., 1998; Zhu et al., 2000). The expression of TGF-β1 is upregulated in hippocampal neurons in ischemia (Zhu et al., 2000) and TGF-β1 mRNA increases in astrocyte cultures exposed to IFN-γ, LPS or TNF-α (Wesselingh et al., 1990).

In the central nervous system, TGF-β1 has a direct neuroprotective effect (Henrich-Noack at al., 1996). Both TGF-β1 and NGF seem to protect neurons against apoptosis by activating the production of cell survival-promoting proteins like members of the Bcl-2 family and proteins that participate in calcium homeostasis (Mattson, 2000; Zhu et al., 2004a). TGF-β1 also has an indirect neuroprotective effect by preventing overactivation of microglial cells by modulating several glial functions (Chen et al., 2002; Mittaud et al., 2002; De Sampaio et al., 2002, Eyüpoglu et al., 2003). TGF-β1 secreted by hippocampal cells modulates the production of NO and super oxide radicals by microglial cells (Herrera-Molina and von Bernhardi, 2005), and inhibits:(1) the production of the pro-inflammatory molecules IL-1 and TNF-α, (2) the induction of NOS, decreasing NO release (Vincent et al., 1997; Mc-Cartney-Francis and Wahl, 2002; Lieb et al., 2003), and (3) IFN-γ production by mononuclear cells (Mouri et al., 2002). It has differential effects in combination with other anti-inflammatory cytokines (Ledeboer et al., 2000). On the other hand, TGF-β1 is also associated with pro-inflammatory events and the potentiation of neurotoxicity (Brown, 1999). Recently, the existence of a reciprocal regulatory interaction between TGF-β1 and IFN-γ has been proposed (Ishida et al., 2004). This is consistent with the deregulation of IFN-γ signaling pathways in TGF-β1 null mice (McCarney-Francis and Wahl, 2002).

Both anti-inflammatory cytokines, like TGF-β, and pro-inflammatory cytokines, such as IL-1β and TNF-α, modulate the inflammatory response of microglia (Figure 6). TGF-β1 and IL-1β, but not TNF-α, decrease NO production. TGF-β1 exerts its biological functions mainly through receptor-activated Smad signaling. In contrast, neuroprotection by TGF-β1 seems to be mediated through activation of PI3 kinase/Akt and MAPK/ERK pathways (Zhu et al., 2002; 2004).

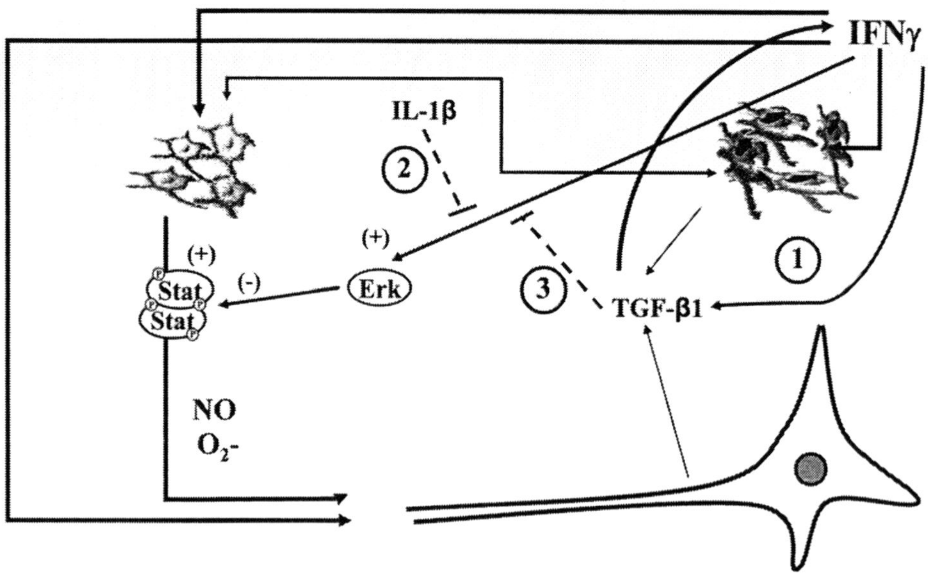

1. IFNγ also induce production of cytokines, IL-1β at early times and later TGFβ. IFNγ and TGFβ show several reciprocal modulatory effects. The increased secretion of TGFβ modulates the production of radical species.
2. IL-1β inhibited the activation of the ERK pathway induced by IFNγ. The effect could be part of an early mechanism to generate an auto limited inflammatory response. If the original stimuli is mild enough it could be sufficient to end glial cell activation.
3. If the inflammatory stimulus persists, the increased secretion of TGFβ in response to the IFNγ will further inhibit the activation of ERK pathway.

Figure 6. Modulation of Microglial Activation by TGFβ and IL1β. IFNγ (or other pro-inflammatory cytokines) activate both glial cells and neurons. Microglial cells respond to IFNγ stimulation by secreting radical species and a series of reactive proteins that also could mediate cell damage. Those effects are mediated at early times by the activation of JAK/STAT as the effector pathway and ERK. The activation of ERK by IFNγ and its phosphorylation of STAT1 in the serine position potentiate the activation of STAT.

The ERK pathway is stimulated by various elements including growth factors, tumor promoters and cytokines. The ERK pathway mediates the activation of iNOS in response to LPS and/or IFN-γ in macrophages (Blanchette at al., 2003) and microglial cells (Han et al., 2002; Pyo et al., 1998). Although STAT1α is the main transcription factor involved in the induction of iNOS, there is an important role of ERK in the stable phosphorylation of STAT1α. We found that ERK transduction pathway is involved in the modulation of oxidative molecules by anti-inflammatory cytokines (TGF-β1) and pro-inflammatory cytokines (IL-1β). IL-1β and TGF-β1 inhibit ERK with different kinetics; transient for IL-1β, at minutes of activation with IFN-γ (30 min) and prolonged for TGF-β1, which inhibits ERK phosphorylation only after hours of activation (24 h) and persists for a long time (Saud et al., 2005). TGF-β1 and IL-1β modulate ERK in a manner reminiscent of a common pathway regulating the amplitude and duration of glial activation in response to various cytokines. The inhibition of microglial activation by pro-inflammatory cytokines suggests that the timing of the activation of the different pathways plays a key role in determining the cellular response. TNF-α, another pro-inflammatory cytokine, does not inhibit NO production or ERK phosphorylation (Saud et al., 2005). IL-1β, induced early after injury, has an autocrine effect (Mrak et al. 1995) and can mediate the auto regulation of the activation of microglial cells.

VIII. Receptors Mediating the Interaction of Glial Cells with Aβ

Aβ interacts with cell membranes (Terzi et al., 1997). The binding of Aβ to the plasma membrane is relevant for understanding its neurotoxicity and the activation of an inflammatory response mediated by glial cells. Aβ binds to numerous molecules, including proteins, proteoglycans and lipids. Membrane lipid domains have various functions including signal transduction, metabolism and cell growth, all of which are involved in changes related to inflammation. Aβ binding leads to both membrane changes (Kremer et al., 2000) and an increase in Aβ fibrillogenesis (Kakio et al., 2002; 2003), two effects associated with cholesterol content and its distribution on the membrane (Hartmann, 2001; Yip et al., 2001).

Several membrane proteins bind the various forms of Aβ and, depending on the receptor; such binding has various effects on neurons and glial cells. It is generally accepted that the aggregated forms show an increased binding. Soluble Aβ binds to the serpin-enzyme complex receptor (SEC-R), which could mediate its endocytosis (Boland et al., 1995), and the insulin receptor (Xie et al., 2002). Both soluble and fibrillar forms of Aβ bind to the α-7-nicotinic receptor for acetylcholine (α-7-nAChR), and could be involved in the cognitive deficits observed in AD (Tran et al., 2002; Wang et al., 2000). Binding of the various forms of Aβ to integrins could influence the regulation of adhesion, internalization and degradation of Aβ (Matter et al., 1998; Bi et al., 2002). The various forms also bind (1) to the receptor for advanced glycosilation end-products (RAGE), inducing oxidative stress and production of cytokines (Lue et al., 2001b, (2) to the SR macrophage receptor with collagenous structure (SR-MARCO), a member of the SR-A family (Re et al., 2002, Alarcón et al., 2005) and (3) to formyl peptide receptor-like 1 (FPRL1) expressed by astrocytes and other cells. FPRL1 mediates a pro-inflammatory response (Tiffany et al., 2001) and is also involved in the uptake and aggregation of Aβ into fibrills by macrophages (Cui et al., 2002). Fibrillar Aβ binds to APP and could be associated with oxidative stress (Lorenzo et al., 2000). Binding to the p75 neurotrophin receptor is associated with both neuronal damage (Yaar et al., 1997; 2002) and protection (Zhang et al., 2003). Fibrillar Aβ also binds to SR-A, SRB, (Husemann et al., 2002) and CD36 (Coraci et al., 2002), whose activation is associated with an inflammatory response. Binding to low density lipoprotein (LDL) receptor-related protein (LRP) could have a role in the clearance mechanisms of Aβ (Arelin et al., 2002), but only when Aβ is associated with ApoE, ApoJ or α2-macroglobulin. LRP level in the brain cortex of AD patients is increased by over 80% (Qiu et al., 2001).

There is mounting evidence that various Aβ oligomers can impair synaptic function in vivo, although the receptors mediating these adverse effects have yet to be identified. Some of the membrane proteins just discussed above could be involved by the activation of second messenger signaling pathways (Dickey et al., 2003; Vitolo et al., 2002; Wang et al., 2004). However, it is unknown whether Aβ only binds to one or a few specific neuronal receptors or nonspecifically affects several receptors or other membrane proteins required for signaling. The binding of Aβ to various members of the SR family can be especially relevant for understanding the association of inflammation with AD (Husemann et al., 2002). Aβ binding is related to the upregulation of expression of some of the SR, the secretion of chemokines and pro-inflammatory cytokines like IL-1β, induction of respiratory burst with increased production of ROS and the activation of various signaling transduction pathways.

Microglial cells express various receptors that bind fibrillar Aβ (fAβ) under normal and pathological conditions. SR are among the receptors mediating the binding and uptake of Aβ *in vitro*. SR include: SR type A (SR-A) (Krieger and Hertz 1994), type B (SR-BI) (Husemann et al., 2002), CD36 (Coraci et al., 2002), RAGE (Yan et al., 1996), LRP (Wolf et al., 1992), mannose receptor (Marzolo et al., 1999) and SR-MARCO. The receptor that mediates Aβ activation of glial cells is unknown. Murine neonatal microglia in culture express SR-A, SR-BI and CD36 (Paresce et al., 1996), while adult mouse microglia only express low levels of CD36 (Husemann et al., 2002) and do not express detectable levels of SR-A (Bell et al., 1994). The expression of SR also changes under certain pathophysiological conditions. SR-A is expressed by activated microglia in the vicinity of senile plaques (Christie et al., 1996).

Astrocytes also express receptors that bind Aβ. The functional role of the adhesion of astrocytes to Aβ is unresolved. While microglia bind to, internalize, and can degrade Aβ, little is known regarding the participation of astrocytes, although astrocytes can phagocytose (Bechmann and Nitsch. 1997). It has been suggested that in contrast with microglia, astrocytes can take up and degrade Aβ without additional stimuli such as cytokines (Wyss-Coray et al., 2003). Unlike microglia, astrocytes only express SR-BI (Husemann and Silverstein. 2001; Alarcón et al., 2005), mannose receptor (Burudi et al., 1999) and SR-MARCO. Astrocytes from normal adult mouse and human brains and from AD brains express SR-BI (Husemann and Silverstein. 2001) and SR-MARCO is expressed in the brains of neonatal and adult rats. Astrocytes expressing SR-MARCO are mostly GFAP-positive, suggesting that SR-MARCO is mainly expressed by activated astrocytes (Alarcón et al., 2005). These observations, and the presence of Aβ in astrocytes surrounding amyloid plaques (Wegiel et al., 2000), led us to propose that dysregulation of Aβ clearance by astrocytes participate in its accumulation in AD. On the other hand, Aβ could bind to astrocytes, activating signal transduction pathways involved in the inflammatory response (discussed below).

fAβ adheres to and is taken up by human and rodent macrophages and microglia (El Khoury et al., 1996; Paresce et al., 1996); both of these cells secrete ROS. Non fibrillar oligomeric $A\beta_{1-42}$ (nfAβ) triggers glial activation, which can lead to the production of potentially toxic molecules such as inflammatory cytokines (IL-1β) and NO (Hu et al., 1998). The adhesion of neonatal astrocytes to fAβ and nfAβ is mediated by a SR that differs from SR-BI (Gu et al., 2000). On the other hand, SR-A is not required for fAβ –induced microglial activation (Antic et al., 2000). Microglia obtained from SR-A knockout newborn mice-expressing SR-BI (Husemann et al., 2001), were only partially blocked from adhering to Aβ with anti-SR-BI, whereas fucoidan totally blocked adhesion of SR-A. Astrocytes and microglia adhere in a concentration-dependent manner to surfaces coated with fibrillar- and non-fibrillar-Aβ, a reaction mediated by a fucoidan- and poly-I-sensitive receptor that is inhibited by immune-neutralization of SR-MARCO. Those observations suggest that SR-MARCO is able to mediate the adhesion of astrocytes and microglia to Aβ (Alarcón et al., 2005).

The binding of Aβ to other SRs can also induce potentially harmful responses. RAGE is a multi-ligand receptor. Its name derives from its first recognized ligand, advanced glycation end-products. The production of RAGE is increased by hyperglycemia, increasing age, and at sites of oxidative stress and inflammation. Binding of RAGE leads to the expression of several pro-inflammatory molecules mediated by the activation of NFkB transcription factor (Hofmann et al., 1999) and leads to the induction of its expression. Such a positive feedback can contribute to chronic inflammation and tissue injury. Immunoreactivity for advanced glycation end products is increased 3-fold in AD compared with control individuals (Vitek et al., 1994). RAGE is expressed by neurons and microglial cells in the brain and both cell types can bind Aβ (Yan et al., 1996). Binding of Aβ to RAGE activates microglial cell. Upon binding to neurons, it induces the production of ROS and the secretion of macrophage-colony stimulating factor (M-CSF) through NFkB. NFkB is also activated by pro-inflammatory cytokines such as IL-1β and TNF-α and is critical for microglia-mediated neurotoxicity induced by Aβ (Chen et al., 2005). M-CSF induces microglial cell migration and proliferation (Du Yan et al., 1997). However, competition assays with ligands for RAGE do not modify the Aβ response suggesting that other receptors also mediate Aβ neurotoxicity (Liu et al., 1997). Binding of Aβ to CD36 activates microglia to secrete ROS, pro-inflammatory cytokines such as IL-1β and TNF-α and several chemokines active on microglial cells, which amplify the inflammatory response (El Khoury et al., 2003). fAβ activates a CD36-dependent signaling cascade involving Src kinase family members, the inflammatory mediators Lyn and Fyn and the mitogen-activated protein kinase p44/42 (El Khoury et al., 2003).

The interaction of the various SRs with Aβ could be complementary and mostly associated with the induction of an inflammatory response. The study of signal transduction pathways activated by the binding of Aβ to SRs should help clarify why inflammation is associated with the activation of SRs. Nothing is known regarding SR-MARCO signaling. However, PKC inhibitors reduce myelin phagocytosis mediated by the related SR-A and complement receptor (Makranz et al., 2004) and ligand binding to SR-A stimulates activation of PI3-kinase and protein-tyrosine kinase and MAPK pathways (Hsu et al., 2001; Kim et al., 2003). Binding to SR-A also increases pro-inflammatory cytokines such as IL-1β and TNF-α (Hsu et al., 2001; Kim et al., 2003), both associated with neurotoxicity in degenerative diseases.

IX. A Proposal: Glial Cell Dysregulation

Aging of the CNS is associated with a pro-inflammatory status that induces functional changes of microglia and astrocytes which can lead to regulatory impairment. This represents a shift in the basal cell reactivity of brain defense responses that can favor progression of neurodegenerative processes leading to AD. There is compelling evidence that the cellular changes responsible for the development of AD are complex in nature and extend beyond the generation of amyloid plaques and neurofibrillary tangles. We propose that Aβ accumulation is a consequence and not the cause on AD pathogenesis. The "inflammation hypothesis" stresses that hyperactive microglia are the primary cause of AD-associated neurotoxicity. In contrast, we propose that AD is not caused by hyperactive but rather dysfunctional microglia (Figure 7). Active microglial cells are needed as scavenger cells in the CNS. However, by not responding to normal regulatory feedback mechanisms and/or having an impairment in their ability to clear Aβ (and other compounds), microglial cells lose their ability to handle potentially toxic compounds and become cytotoxic. Thus microglial cells no longer provide such beneficial actions as removing Aβ (Paresce et al., 1997), eliminating debris, secreting trophic factors and reducing the possibility of further damage by reducing the release of toxic factors. Other changes relevant to AD also occur during aging. Transient hypoxia can lead to mitochondrial dysfunction, resulting in oxidative stress, impairment of membrane integrity and amyloidogenic APP cleavage with a pattern similar to that observed in sporadic AD (Chen et al., 2003).

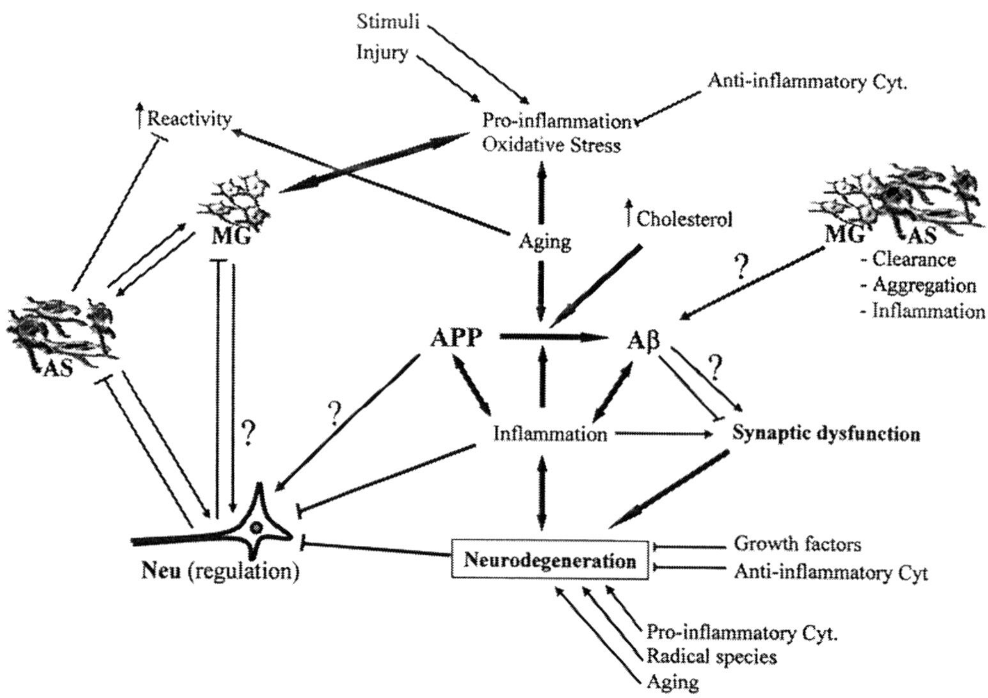

Figure 7. Putting the pieces together: Several different pathogenic events contribute to neurodegeneration in Alzheimer's disease. Inflammatory events, either secondary to aging or to various forms of stimuli or injury, can be responsible of abnormal glial activation, increased production of Aβ and neuronal dysfunction. At late stages, they are also involved directly or indirectly in the induction of neurodegeneration. Glial activation results in the release of various cytokines and the generation of reactive species, contributing to the inflammatory process. The increased expression of Aβ results in further pro-inflammatory activation, perpetuating the glial activation, synaptic dysfunction and cell damage. Inflammatory mediators, neurodegenerative processes and the trophic effect of APP and Aβ also affect neuronal regulation, impairing their modulation of glial activation.

A state of increased inflammatory reactivity potentiates the response of glial cells and can result in a pathological overreaction of microglia when exposed to a condition that normally would fail to induce their activation such as Aβ, a compound that is not a potent activator by itself (von Bernhardi and Eugenín., 2004). Also, abnormal glial activation can impair the capacity of these cells to uptake and degrade Aβ (Rogers et al., 2002). Normal subjects and AD patients have similar concentrations of soluble Aβ in the cerebrospinal fluid (Seubert et al., 1992) and Aβ expression by cells is constitutive at least in culture (Haass et al., 1992). Impairment of uptake or degradation would allow Aβ accumulation by reducing its rate of degradation even in the absence of changes of Aβ production. The overproduction or decreased degradation of Aβ is just one part of a series of detrimental changes associated with aging. Oxidative stress and the generation of free radicals can also promote the processing of APP to the amyloidogenic pathway. The persistence of Aβ (or other potentially toxic compounds) can establish a vicious circle, further potentiating an inflammatory reaction. In time, chronically or multiple-event activated microglia and astrocytes would become neurotoxic by the release of inflammatory cytokines, proteolytic enzymes, complement factors and reactive intermediaries.

The increase in TGF-β1 production in response to pro-inflammatory conditions (Ramírez et al., 2005) is one of the regulatory mechanisms secondary to cell activation constituting a feedback loop (Herrera-Molina and von Bernhardi, 2005). TGF-β1 deactivates microglial cells and abolishes neurotoxicity (Eyüpoglu et. al., 2003). The induction of TGF-β1 by pro-inflammatory molecules would limit the temporal and spatial extent of the inflammatory response. Its modulatory effect involves the activation of MAPK- ERK pathways (Saud et al., 2005), which seems to be neuroprotective (Zhu et al., 2002 and 2004). The MAPK pathways coordinate multiple cellular pathways involving signals such as inflammatory cytokines, growth factors, substrate- and cell-adhesion molecule receptors. They can be key elements for cellular integrity and handling of inflammation. In the context of AD, it should prove useful to better understand the effect of chronic inflammation on the activation of MAPK.

The regional and cellular differences in cytotoxicity of microglial cells probably depend on multiple factors related to cell identity and their environment. Thus, regulation of glial activation can be impaired under sustained pro-inflammatory conditions as in aging. The activation of glial cells by Aβ is low, but it is markedly amplified by the inflammatory priming of glial cells. Also, microglial reactivity to Aβ is attenuated by the presence of astrocytes. However, attenuation is not observed when glial cells are exposed to pro-inflammatory factors (von Bernhardi and Eugenín, 2004), again suggesting that pro-inflammatory conditions can be determining factors in the susceptibility of microglia to modulation. The inflammatory priming of glial cells, rather than Aβ aggregates, could be the principal trigger for glial activation and inflammation in AD. In the brain, Aβ interacts with both microglia and astrocytes and under physiological conditions the microglial reaction to Aβ would be mild and short lasting and down regulated by the activation of astrocytes. However, under pro-inflammatory conditions, down regulation by astrocytes could fail thus enhancing microglial activation and secretion of inflammatory cytokines. This in turn can increase cytotoxicity and Aβ aggregation, establishing a vicious circle. In early onset AD, enhancement of microglial activation or impairment of their regulatory signals could be observed at earlier ages because of the accumulation of more frequent or stronger stimuli upon glial cells, as proposed for the pathological aging (Figure 3). Even in familial AD the production of increased amounts of Aβ can become a stimulus for the activation of glial cells, triggering an inflammatory response. In these examples, dementia is established at younger ages, but the neuropathological landmarks become more abundant as the patient ages, suggesting that the association to aging is maintained.

The various observations discussed above support our hypothesis that glial dysregulation leads to cytotoxicity and the neurodegenerative process characteristic of AD. Determination of the receptors mediating and regulating microglial activation and understanding the molecular mechanisms activated by these receptors are needed to generate better ways to treat and prevent neurodegenerative diseases. Because SR interactions could mediate activation of astrocytes and microglial cells, the signaling pathways associated with SR binding should be elucidated. The binding of Aβ to different SRs could induce various cellular responses depending on the activation states of intracellular signaling pathways.

Conclusion

I propose that increased production/levels and deposition of Aβ in sporadic AD occurs secondary to inflammatory or oxidative stress changes associated with aging. There is *in vivo* evidence that the deposition of Aβ occurs after oxidative stress (Nunomura et al., 2001) or secondary to neuronal stress and its aggregation is increased in pro-inflammatory conditions. Evidence suggests that Aβ has several potential deleterious effects. However, Aβ can have beneficial effects at physiological concentrations and become toxic when its production or levels is/are increased resulting in pathology. The increased generation of Aβ and damage in AD can depend on the initial oxidative stress or inflammatory event(s) (e.g., as observed in aging). This event(s) results in increased production and processing of APP to Aβ and modulates the interaction of Aβ with glial cells, impairing its clearance and potentiating a reactive inflammatory response.

Acknowledgements

I thank Dr. Howard Etlinger for his longstanding support, his always insightful suggestions and his critical reading of the manuscript. My gratitude for the good work of my students and the staff members of my laboratory, especially G. Ramírez, R. Alarcón and R. Herrera-Molina. The author's work is supported by grant 1040831 from FONDECYT to RvB.

REFERENCES

Abramov, A.Y., Canevari, L., Duchen. M.R. (2003). Changes in intracellular calcium and glutathione in astrocytes as the primary mechanism of amyloid neurotoxicity. *J. Neurosci.* 23, 5088-5095.

Aguado, F., Ballabriga, J., Pozas, E., Ferrer, I. (1998). TrkA immunoreactivity in reactive astrocytes in human neurodegenerative diseases and colchicines-treated rats. *Acta Neuropathol.* (Berlin) 96, 495-501.

Akama, K.T., Albanese, C., Pestell, R.G., Van Eldik, C.J. (1998). Amyloid beta-peptide stimulates nitric oxide production in astrocytes through an NFkappaB-dependent mechanism, *PNAS USA.* 95, 5795-5800.

Akiyama, H., Barger, S., Barnum, S., Bradt, B., Bauer, J., Cle, G.M., Cooper, N.R., Eikelenboom, P., Emmerling, M., Fiebich, B.L., Finch, C.E., Frautschy, S., Griffin, W.S.T., Hampel, H., Hull, M., Landreth, G., Lue, L., Mrak, R., Mackenzie, I.R., McGeer, P., O'Banion, M.K., Pachter, J., Pasinetti, G., Plata-Salaman, C., Rogers, J., Rydel, R., Shen, Y., Streit, W., Strohmeyer, R., Tooyoma, I., Van Muiswinkel, F., Veerhuis, R., Walker, S., Wegrzyniak, B., Wenk, G., Wyss-Coray, T. (2000). Inflammation and Alzheimer's disease. *Neurobiol. Aging* 21, 383-421.

Alarcón, R., Fuenzalida, C., Santibañez, M. von Bernhardi, R. (2005). Expression of Scavenger Receptors in glial cells: Comparing the adhesion of astrocytes and microglia from neonatal rats to surface-bound β-amyloid. *JBC.* 280, 30406-30415.

Aleman, A., Verhaar, H.J., De Haan, E.H., De Vries, W.R., Samson, M.M., Drent, M.L., Van der Veen, E.A., Kopeschaar, H.P. (1999). Insulin-like growth factor-I and cognitive function in healthy older men. *J. Clin. Endocrinol. Metab.* 84, 471-475.

Allison, A.C., Cacabelos, R., Lombardi, V.R., Alvarez, X.A., Vigo, C. (2001). Celastrol, a potent antioxidant and anti-inflammatory drug, as a possible treatment for Alzheimer's disease. Prog. Neuro-Psychopharmacol. *Biol. Psychiatry* 25, 1341-1357.

Antic, A., Dzenko, K.A., Pachter, J.S. (2000). Engagement of the scavenger receptor is not responsible for beta-amyloid stimulation of monocytes to a neurocytopathic state. *Exp. Neurol.* 161, 96-101.

Apelt, J., Schliebs, R. (2001). Beta-amyloid-induced glial expression of both pro- and anti-inflammatory cytokines in cerebral cortex of aged transgenic Tg2576 mice with Alzheimer plaque pathology. *Brain Res.* 894, 21-30.

Arelin, K., Kinoshita, A., Whelan, C.M., Irizarry, M.C., Rebeck, G.W., Strickland, D.K., Hyman, B.T. (2002). LRP and senile plaques in Alzheimer's disease: colocalization with apolipoprotein E and with activated astrocytes. *Brain Res. Mol. Brain Res.* 104, 38-46.

Arnett, H.A., Masson, J., Marino, M., Suzuki, K., Matsushima, G.K., Ting, J.P.Y. (2001). TNFα promotes proliferation of oligodendrocyte progenitors and remyelinization. *Nature Neurosci.* 4, 1116-1121.

Arnett, H.A., Wang, Y., Matsushima, G.K., Suzuki, K., Ting, J.P.Y. (2003). Functional genomic analysis of remyelination reveals importance of inflammation in oligodendrocytes regeneration. *J. Neurosci.* 23, 9824-9832.

Arnold, J., Dawson, S., Fergusson, J., Lowe, J., Landon M., Mayer, J. (1998). Ubiquitin and its role in neurodegeneration. In: Van Leeuwen F.W., Salehi, A., Giger, R.J., Holtmaat,

A.J.G.D., Verhagen, J. (Eds.), *Progress in Brain Research*, Elsevier, Amsterdam, vol 117, pp. 23-34.

Avital, A., Goshen, I., Kamsler, A., Segal, M., Iverfeldt, K., Richter-Levin, G., Yirmiya, R. (2003). Impaired interleukin-1 signaling is associated with deficits in hippocampal memory processes and neural plasticity. *Hippocampus* 13, 826-834.

Bales, K., Du, Y., Dodel, R.C., Yan, G-M., Hamilton-Byrd, E., Paul, S.M. (1998). The NF-kappaB/Rel family of proteins mediates Abeta-induced neurotoxicity and glial activation, *Molec. Brain Res.* 57, 63-72.

Bal-Price, A., Brown, C. (2001). Inflammatory neurodegeneration mediated by nitric oxide from activated glia-inhibiting neuronal respiration, causing glutamate release and excitotoxicity. *J. Neurosci.* 17, 6480-6491.

Baltrons, M.A., Pedraza, C., Heneka, M., García, A. (2002). β-amyloid peptides decreased soluble guanylyl cyclase expression in astroglial cells. *Neurobiol. Dis.* 10, 139-149.

Bard, F., Cannon, C., Barbour, R., Burke, R.L., Games, D., Grajeda, H., Guido, T., Hu, K., Huang, J., Johnson-Wood, K., Khan, K., Kholodenko, D., Lee, M., Lieberburg, I., Motter, R., Nguyen, M., Soriano, F., Vasquez, N., Weiss, K., Welch, B.. Seubert, P., Schenk, D., Yednock, T. (2000). Peripherally administered antibodies against amyloid beta-peptide enter the central nervous system and reduce pathology in a mouse model of Alzheimer's disease. *Nat. Med.* 6, 916-919.

Barger, S., Hörster, D., Furukawa, K., Goddman, Y., Krieglstein, J., Mattson, M. (1995). Tumor necrosis factors alpha and beta protect against APP toxicity: evidence for involvement of a kB-binding factor and attenuation of peroxide and Ca^{+2} accumulation. *Proc. Natl. Acad. Sci. USA.* 92, 9328-9332.

Barger, S.W., Harmon, A.D. (1997). Microglial activation by Alzheimer amyloid precursor protein and modulation by apolipoprotein E. *Nature* 388, 878-881.

Baron, P., Galimberti, D., Meda, L., Prat, E., Scarpini, E., Conti, G., Moggio, M., Prelle, A., Scarlato, G. (2000). Synergistic effect of β-amyloid protein and interferon gamma on nitric oxide production by C2C12 muscle cells. *Brain.* 123, 374-379.

Barone, F., Arvin, B., White, B., Miller, A., Webb, C., Willette, R., Lysko, P., Feuerstein, G. (1997). Tumor necrosis factor alpha. A mediator of focal ischemic brain injury. *Stroke* 28, 1233-1244.

Barron, K.D. (1995). The microglial cell. A historical review, *J. Neurol. Sci.* 134, 57-68.

Bartus, R.T. (2000). On degenerative diseases, models, and treatment strategies: Lessons learned and lessons forgotten a generation following the cholinergic hypothesis. *Exp. Neurol.* 163, 495-520.

Basu, A., Krady, J., Levinson, S. (2004). Interleukin-1: A master regulator of neuroinflammation. *J. Neurosci. Res.* 78, 151-156.

Bechmann, I., Nitsch, R. (1997). Identification of phagocytic glial cells after lesion-induced anterograde degeneration using double-fluorescence labeling: combination of axonal tracing and lectin or immunostaining. Histochem. *Cell Biol.* 107, 391-397,

Bell, M.D., Lopez-Gonzalez, R., Lawson, L., Hughes, D., Fraser, I., Gordon, S., Perry, V.H. (1994). Upregulation of the macrophage scavenger receptor in response to different forms of injury in the CNS. *J. Neurocytol.* 23, 605-613,

Benveniste, E.N. (1998). Cytokine actions in the central nervous system. *Cytokine and Growth Factors Rev.* 9, 259-275.

Benveniste, E., Nguyen, V., O'Keefe, G. (2001). Immunological aspects of microglia: relevance to Alzheimer's disease. *Neurochem. Int.* 39, 381-391.

Benzing, W.C., Wujek, J.R., Ward, E.K., Shaffer, D., Ashe, K.H., Youkin, S.G., Brunden, K.R. (1999). Evidence for glial-mediated inflammation in aged APP(SW) transgenic mice. *Neurobiol. Aging* 20, 581-589.

Bertoni-Freddari, C., Fattoretti, P., Casoli, T., Caselli, U., Meier-Ruge, W. (1996). Deterioration threshold of synaptic morphology in aging and senile dementia of Alzheimer's type. *Anal. Quant. Cytol. Histol.* 18, 209-213.

Bi, X., Gall, C.M., Zhou, J., Lynch, G. (2002). Uptake and pathogenic effects of amyloid beta peptide 1-42 are enhanced by integrin antagonists and blocked by NMDA receptor antagonists. *Neuroscience* 112, 827-840.

Blanchette, J., Jaramillo, M., Olivier, M. (2003). Signalling events involved in interferon-β-inducible macrophage nitric oxide generation. *Immunology* 108, 513-522.

Blasko, I., Marx, F., Steiner, E., Hartmann, T., Grubeck-Loebenstein, B. (1999). TNFalpha plus IFNgamma induce the production of Alzheimer beta amyloid peptides and decrease the secretion of APPs. *FASEB J.* 13, 63-68.

Blasko, I., Apochal, A., Boeck, G., Hartmann, T., Grubeck-Loebenstein, B., Ransmayr, G. (2001). Ibuprofen decreases cytokine-induced amyloid Beta production in neuronal cells. *Neurobiol. Dis.* 8, 1094-1101.

Blasko, I., Stampfer-Kountchev, M., Robatscher, P., Veerhuis, R., Eikelenboom, P., Grubeck-Loebenstein, B. (2004). How chronic inflammation can affect the brain and support the development of Alzheimer's disease in old age: the role of microglia and astrocytes. *Aging cell* 3, 169-176.

Blondel, O., Collin, C., McCarran, W.J., Zhu, S., Zamostiano, R., Gozes, I., Brenneman, D.E., McKay, R.D. (2000). A glia-derived signal regulating neuronal differentiation. *J. Neurosci.* 20, 8012-8020.

Bodovitz, S., Klein, W.L. (1996). Cholesterol modulates alpha-secretase cleavage of amyloid precursor protein. *J. Biol. Chem.* 271, 4436-4440.

Boland, K., Manias, K., Perlmutter, D.H. (1995). Specificity in recognition of amyloid-beta peptide by the serpin-enzyme complex receptor in hepatoma cells and neuronal cells. *J. Biol. Chem.* 270, 28022-28028.

Bowman, C.C., Rasley, A., Tranguch, S.L., Mattiott. (2003). Cultured astrocytes express toll-like receptors for bacterial products. *Glia* 43, 281-291.

Braak, H., Braak, E., Bohl, J. (1993). Staging of Alzheimer-related cortical destruction. *Eur. Neurol.* 33, 403-408.

Braak, H., Braak, E. (1997). Frequency of status of Alzheimer-related lesions in different age categories. *Neurobiol. Aging.* 18, 351-357.

Broe, G.A., Grayson, D.A., Creasey, H.M., Waite, L.M., Casey, B.J., Bennett, H.P., Brooks, W.S., Halliday, G.M. (2000). Anti-inflammatory drugs protect against Alzheimer's disease at low doses. *Arch. Neurol.* 57, 1586-1591.

Brown, D.R. (1999). Dependence of neurones on astrocytes in a co-culture system renders neurones sensitive to transforming growth factor β1-induced glutamate toxicity. *J. Neurochem.* 72, 943-953.

Bsibi, M., Ravid, R., Gveric, D., Noort, J.M. (2002). Broad expression of Toll-like receptors in the human central nervous system. J. Neuropathol. *Exp. Neurol.* 61, 1013-1021.

Burudi, E.M., Riese, S., Stahl, P.D., Regnier-Vigouroux, A. (1999). Identification and functional characterization of the mannose receptor in astrocytes. *Glia* 25, 44-55.

Busciglio, J., Gabuzda, D.H., Matsudaira, P., Yankner, B.A. (1993). Generation of beta-amyloid in the secretory pathway in neuronal and nonneuronal cells. *Proc. Natl. Acad. Sci. USA* 90, 2092-2096.

Butterfield, D.A., Castegna, A., Lauderback, C.M., Drake, J. (2002). Evidence that amyloid beta-peptide-induced lipid peroxidation and its sequelae in Alzheimer's disease brain contribute to neuronal death. *Neurobiol. Aging* 23, 655-664.

Bye, N., Zieba, M., Wreford, N.G., Nichols, N.R. (2001). Resistance of the dentate gyrus to induced apoptosis during ageing is associated with increases in transforming growth factor-beta 1 messenger RNA. *Neuroscience* 105, 853-862.

Cagnin, A., Brooks, D.J., Kennedy, A.M., Gunn, R.N., Myers, R., Rurkheimer, F.E., Jones, T., Banati, R.B. (2001). In-vivo measurement of activated microglia in dementia. *Lancet* 358, 461-467.

Canevari, L., Abramov, A.Y., Duchen, M.R. (2004). Toxicity of amyloid beta peptide: tales of calcium, mitochondria, and oxidative stress. *Neurochem. Res.* 29, 637-650.

Chalazonitis, A., Kalberg, J., Twardzik, D.R., Morrison, R.S., Kessler, J.A. (1992). Transforming growth factor β has neurotrophic actions on sensory neurons in vitro and is synergistic with nerve growth factor. *Dev. Biol.* 152, 121-132.

Chang, J.Y., Martin, D.P., Johnson Jr., E.M. (1990). Interferon suppresses sympathetic neuronal cell death caused by nerve growth factor deprivation. *J. Neurochem.* 55, 436-445.

Chao, C.C., Hu, S., Peterson, P.K. (1995). Modulation of human microglial cell superoxide production by cytokines. *J. Leukoc. Biol.* 58, 35-70.

Chao, C.C. (1996). Cytokine-stimulated astrocytes damage human neurons via a NO mechanism. *Glia* 16, 276-284.

Chapman, P.F., White, G.L., Jones, M.W., Cooper-Blacketer, D., Marshall, V.J., Irizarry, M., Younkin, L., Good, M.A., Bliss, T.V., Hyman, B.T. Youkin, S.G., Hsiao, K.K. (1999). Impaired synaptic plasticity and learning in aged amyloid precursor protein transgenic mice. *Nat. Neurosci.* 2, 271-276.

Chen, S., Luo, D., Strit, W., Harrison, J.K. (2002). TGF-β1 up regulates CX3CR1 expression and inhibits fractalkine-stimulated signaling in rat microglia. *J. Neuroimmunol.* 133, 46-55.

Chen, G.J., Xu, J., Lahousse, S.A., Caggiano, N.L., de la Monte, S.M. (2003). Transient hypoxia causes Alzheimer-type molecular a biochemical abnormalities in cortical neurones: potential strategies for neuroprotection. *J. Alzheimer's Dis.* 5, 209-228.

Chen, J., Zhou, Y., Mueller-Steiner, S., Chen, L-F., Kwon, H., Yi, S., Mucke, L., Gan, Li. (2005). SIRT1 protects against microglia-dependent amyloid-β toxicity through inhibiting NF-kB signaling. *J. Biol. Chem.* 280, 40364-40374.

Cheng, B., Christakos, S., Mattson, M. (1994). Tumor necrosis factors protect neurons against metabolic-excitotoxic insults and promote maintenance of calcium homeostasis. *Neuron* 12, 139-153.

Choi, J., Malakowsky, C.A., Talent, J.M., Conrad, C.C., Carroll, C.A., Weintraub, S.T., Gracy, R.W. (2003). Anti-apoptotic proteins are oxidized by Abeta25-35 in Alzheimer's fibroblasts. *Biochim. Biophys. Acta* 1637: 135-141.

Christie, R.H., Freeman, M., Hyman, B.T. (1996). Expression of the macrophage scavenger receptor, a multifunctional lipoprotein receptor, in microglia associated with senile plaques in Alzheimer's disease. *Am. J. Pathol.* 148, 399-403.

Combs, C.K., Karlo, J.C., Kao, S-C., Landreth, G.E. (2001). beta-Amyloid stimulation of microglia and monocytes results in TNFalpha-dependent expression of inducible nitric oxide synthase and neuronal apoptosis, *J. Neurosci.* 21, 1179-1188.

Coraci, I.S., Husemann, J., Berman, J.W., Hulette, C., Dufour, J.H., Campanella, G.K., Luster, A.D., Silverstein, S.C., El Khoury, J.B. (2002). CD36, a class B scavenger receptor, is expressed on microglia in Alzheirmer's disease brains and can mediate production of reactive oxygen species in response to beta-amyloid fibrills. *Am. J. Pathol.* 160, 101-112.

Coulson, E.J., Paliga, K., Beyreuter, K., Masters, C.L. (2000). What the evolution of the amyloid precursor protein supergene family tells us about its function. *Neurochem Int.* 36, 175-184.

Cui, Y., Le, Y., Yazawa, H., Gong, W., Wang, J.M. (2002). Potential role of the formyl peptide receptor-like 1 (FPRL1) in inflammatory aspects of Alzheimer's disease. *J. Leukoc. Biol.*. 72, 628-635.

Cunningham, Deacon, R., Wells, H., Boche, D., Waters, S., Diniz, C.P., Scott, H., Rawlins, J.N., Perry, VH. (2003). Synaptic changes characterize early behavioural signs in the ME7 model of murine prion disease. *Eur. J. Neurosci.* 17, 2147-2155.

D'Andrea, M.R., Cole, GM., Ard, M.D. (2004). The microglial phagocytic role with specific plaque types in the Alzheimer disease brain. *Neurobiol. Aging* 25, 675-683.

Davies, C.A., Mann, D.M., Sumpter, P.Q., Yates, P.O. (1987). A quantitative morphometric analysis of the neuronal and synaptic content of the frontal and temporal cortex in patients with Alzheimer's disease. *J. Neurol. Sci.* 78, 151-164.

Davis, D.G., Schmitt, F. A., Wekstein, D.R. Markesbery, W.R. (1999). Alzheimer neuropathologic alterations in aged cognitively normal subjects. *J. Neurpathol. Exp. Neurol.* 55, 376-388.

DeKosky, S.T., Scheff, S.W. (1990). Synapse loss in frontal cortex biopsies in Alzheimer's disease: correlation with cognitive severity. *Ann. Neurol.* 27, 457-464.

De Kosky, S.T., Styren, S.D., O'Malley, M.E., Gross, J.R., Kochanek, P., Marion, D., Evans, C.H., Robbins, P.D. (1996). Interleukin-1 receptor antagonist suppresses neurotrophin response in injured rat brain. *Ann. Neurol.* 39, 123-127.

De la Monte, S.M., Chiche, J., von dem Bussche, A., Sanyal, S., Lahousse, S.A., Janssens, S.P., Bloch, K.D. (2003). Nitric oxide synthase-3 overexpression causes apoptosis and impairs neuronal mitochondrial function: relevance to Alzheimer's-type neurodegeneration. *Lab. Invest.* 83, 287-298.

Delanty, N., Vaughan, C.J., Sheehy, N. (2001). Statins and neuroprotection. *Expert Opin. Investig. Drugs.* 10, 1847-1853.

de la Torre, J.C. (2002). Alzheimer's disease as a vascular disorder: nosological evidence. *Stroke* 33, 1152-1162.

Dell'Albani, P., Santangelo, R., Torrisi, L., Nicoletti, V.G., de Vellis, J., Giuffrida Stella A.M. (2001). JAK/STAT signaling pathway mediates cytokine-induced iNOS expression in primary astroglial cell cultures. *J. Neurosci. Res.* 65, 417-424.

De Sampaio, E., Spohr, T.C., Martinez, R., da Silva, E.F., Neto, V.M., Gomez, F.C. (2002). Neuro-Glia interaction effects on GFAP gene: a novel role for TGFβ1. *Eur. J. Neurosci.* 16, 2059-2069.

De Simone, R., Levi, G., Aloisi, F. (1998). Interferon-γ gene expression in rat central nervous system glial cells. *Cytokine* 10, 418-422.

DeWitt, D.A., Perry, G., Cohen, M., Doller, C., and Silver, J. (1998). Astrocytes regulate microglial phagocytosis of senile plaque cores of Alzheimer's disease. *Exp. Neurol.* 149, 329-340

Dhandapani, K.M., Hadman, M., De Sevilla, L., Wade, M.F., Mahesh, V.B., Brann, D.W. (2003). Astrocyte protection of neurons. Role of transforming growth factor-β signaling via c-jun-Ap-1 protective pathway. *J. Biol. Chem.* 278, 43329-43339.

Dickey, C.A., Loring, Z.F., Montgomery, J., Gordom, M.N., Eastman, P.S., Morgan, D. (2003). Selectively reduced expression of synaptic plasticity-related genes in amyloid precursor protein + presenilin-1 transgenic mice. *J. Neurosci.* 23, 5219-5226.

Dickson, D.W. (1997). The pathogenesis of senile plaques. *J. Neuropathol. Exp. Neurol.* 56, 321-339.

Du Yan, S., Zhu, H., Fu, J., Yan, S.F., Roher, A., Tourtel-Iotte, W.W., Rajavashisth, T., Chen, X., Godman, G.C., Stern, D., Schmidt, A.M. (1997). Amyloid-beta peptide-receptor for advanced glycation end product interaction elicits neuronal expression of macrophage-colony stimulating factor: a proinflammatory pathway in Alzheimer's disease. *Proc. Natl. Acad. Sci. USA.* 94, 5296-5301.

Duyckaerts, C., Colle, M.A., Dessi, F., Crignon, Y., Piette, F., Hauw, J.J. (1998). The progression of the lesions in Alzheimer-disease- insights from a prospective clinicopathological study. *J. Neural. Trans. Suppl.* 53, 119-126.

Dziedzic, T., Wybranska, I., Dembinska-Kiec, A., Klimkowicz, A., Slowik, A., Pankiewicz, J., Adzienicka, A., Szczudlik, A. (2003). Dexamethasone inhibits TNF-alpha synthesis more effectively in Alzheimer's disease patients than in healthy individuals. *Dement. Geriatr. Cogn. Disord.* 16, 283-286.

Ebinu, J.O., Yankner, B.A. (2002). A RIP tide in neuronal signal transduction. *Neuron* 34, 499-502.

Eikelenboom, P., Bate, C., Van Gool, W.A., Hoozemans, J.J.M., Rozemuller, J.M., Veerhuis, R., Williams, A. (2002). Neuroinflammation in Alzheimer's disease and Prion disease. *Glia* 40, 232-239.

Eikelenboom, P., van Gool, W.A. (2004). Neuroinflammatory perspectives on the two faces of Alzheimer's disease. *J. Neural. Transm.* 111, 281-294.

El Khoury, J., Hickman, S.E., Thomas, C.A., Cao, L., Silverstein, S.C., Loike, J.D. (1996). Scavenger receptor-mediated adhesion of microglia to beta-amyloid fibrills. *Nature* 382, 716-719.

El Khoury, J.B., Moore, K.J., Means, T.K., Leung, J., Terada, K., Toft, M., Freeman, M.W., Luster, A.D. (2003). CD36 mediates the innate response to beta-amyloid. *J. Exp. Med.* 197, 1657-1666.

Eriksen, J.L., Sagi, S.A., Smith, T.E., Weggen, S., Das, P., McLendon, D.C., Ozols, W., Jessing, K.W., Zavitz, K.H., Koo, E.H., Golde, T.E. (2003). NSAIDs and enantiomers of flurbiprofen target gamma-secretase and lower Abeta 42 in vivo. *J. Clin. Invest.* 112, 440-449.

Etminan, M., Gill, S., Samii, A. (2003). Effect of non-steroidal anti-inflammatory drugs on risk of Alzheimer's disease: systematic review and meta-analysis of observational studies. *BMJ* 327, 128-131.

Eyüpoglu, L.Y., Bechmann, L., Nitsch, R. (2003). Modification of microglial function protects from lesion-induced neuronal alteration and promotes sprouting in the hippocampus. *FASEB J.* 17, 1110-1111.

Farkas, E., De Jong, G.I., de Vos, R.A., Jansen Steur, E.N., Luiten, P.G. (2000). Pathological features of cerebral cortical capillaries are doubled in Alzheimer's disease and Parkinson's disease. *Acta Neuropathol* (Berl). 100, 395-402.

Farrer, L.A., Cupples, L.A., Haines, J.L., Hyman, B., Kukull, W.A., Mayeux, R., Myers, R.H., Pericak-Vance, M.A., Risch, N., van Duijn, C.M. (1997). Effects of age, sex, and ethnicity on the association between apolipoprotein E genotype and Alzheimer's disease. A meta-analysis. APOE and Alzheimer's disease Meta Analysis Consortium. *JAMA* 278, 1349-1356.

Fassbender, K., Simons, M., Bergmann, C., Stroick, M., Luthjohann, D., Keller, P., Runz, H., Kuhl, S., Bergmann, K, von., Hennerici, M., Beyreuther, K., Hartmann, T. (2001). Simvastatin strongly reduces levels of Alzheimer's disease beta-amyloid peptides Abeta 42 and A beta 40 in vitro and in vivo. *Proc. Natl. Acad. Sci. USA* 98, 5856-5861.

Finkel, T. (2000). Redox-dependent signal transduction. *FEBS Lett,* 476, 52-54.

Flanders, K.C., Ren, R.F., Lippa, C.F. (1998). Transforming growth factor-betas in neurodegenerative disease. *Prog. Neurobiol.* 54, 71-85.

Forloni, G., Mangiarotti, F., Angeretti, N., Lucca, E., De Simoni, M.G. (1997). Beta-amyloid fragment potentiates IL-6 and TNF-alpha secretion by LPS in astrocytes but not in microglia, *Cytokine* 9, 759-762.

Friedman, W.J. (2001). Cytokines regulate expression of the type 1 interleukin-1 receptor in rat hippocampal neurons and glia. *Exp. Neurol.* 168, 23-31.

Fu, C., Chute, D.J., Farag, E.S., Garakian, J., Cummings, J.L., Vinters, H.V. (2004). Comorbidity in dementia: an autopsy study. *Arch. Pathol. Lab. Med.* 128, 32-38.

Geula, C., Wu, C.K., Saroff, D., Lorenzo, A., Yuan, M., Yankner, B.A. (1998). Aging renders the brain vulnerable to amyloid beta-protein neurotoxicity. *Nat. Med.* 4, 827-831.

Gilman, S., Koller, M., Black, R.S., Jenkins, L., Griffith, S.G., Fox, N.C., Eisner, L., Kirby, L., Merce Boada, M., Forette, F., Orgogozo, J-M. (2004). Neuropsychological, CSF, and neuropathological effects of A-Beta immunotherapy (AN1792) of Alzheimer's disease in an interrupted trial. *Neurobiol. Aging.* 25, S2, 84.

Giulian, D., Baker, T.J. (1986). Characterization of amoeboid microglia isolated from developing mammalian brain. *J. Neurosci.* 6, 2163-2178.

Giulian, D., Vaca, K., Corpuz, M. (1993). Brain glia release factors with opposing actions upon neuronal survival. *J. Neurosci.* 13, 29-37.

Giulian, D. Corpuz, M. (1993). Microglial secretion products and their impact on the nervous system. *Adv. Neurol.* 59, 315-320.

Giulian, D., Li, J., Li, X., George, J., Rutecki, P. (1994). The impact of microglia-derived cytokine upon gliosis in the CNS. *Dev. Neurosci.* 16, 128-136.

Giulian, D., Haverkamp, L.J., Li, J., Karshin, W.L., Yu, J., Tom, D., Li, X., Kirkpatrick, J.B. (1995). Senile plaques stimulate microglia to release a neurotoxin found in Alzheimer brain, *Neurochem. Int.* 27, 119-137.

Goedert, M. (1998). Neurofibrillary Pathology of Alzheimer's Disease and other taupathies. In: Van Leeuwen F.W., Salehi, A., Giger, R.J., Holtmaat, A.J.G.D., Verhagen, J. (Eds.), *Progress in Brain Research*, Elsevier, Amsterdam, vol 117, pp. 287-306.

Goldgaber, D., Harris, H.W., Hla, T., Maciag, T., Donnelly, R.J., Jacobsen, J.S., Vitek, M.P., Gajdusek, D.C. (1989). Interleukin 1 regulates synthesis of amyloid beta-protein precursor mRNA in human endothelial cells. Proc. Natl. Acad. Sci. USA. 86, 7606-7610.

Goodwin, J.L., Uemura, E., Cunnick, J.E. (1996). Microglial release of nitric oxide by synergistic action of β-amyloid and IFNγ. *Brain Res.* 692, 207-214.

Griffin, W.S., Sheng, J.G., Royston, M.C., Gentleman, S.M., McKenzie, E.J., Graham, D.I., Roberts, G.W., Mrak, R.E. (1998). Glial-neuronal interactions in Alzheimer's disease: the potential role of a "cytokine cycle" in disease progression. *Brain Pathol.* 8, 65-72.

Griffin, W.S., Mrak, R.E. (2002). Interleukin-1 in the genesis and progression of and risk for development of neuronal degeneration in Alzheimer's disease. *J. Leukoc. Biol.* 72, 233-238.

Gu, X., Lawrence, R., Krieger, M. (2000). Dissociation of the high density lipoprotein and low density lipoprotein binding activities of murine scavenger receptor class B type I (mSR-BI) using retrovirus library-based activity dissection. *J. Biol. Chem.* 275, 9120-9230.

Gylys, K.H., Fein, J.A., Yang, F., Willey, D.J., Miller, C.A., Cole, G.M. (2004). Synaptic changes in Alzheimer's disease: increased amyloid-β and gliosis in surviving terminals is accompanied by decreased PSD-95 fluorescence. *Am. J. Pathol.* 165, 1809-1817.

Haass, C., Schlossmacher, M.G., Hung, A.Y., Vigo-Pelfrey, C., Mellon, A., Ostaszewski, B.L., Lieberburg, I., Koo, E.H., Schenk, D., Teplow, D.B., Selkoe, D.J. (1992). Amyloid β-peptide is produced by cultured cells during normal metabolism. *Nature* 359, 322-325.

Haass, C. (2004). Take five-BACE and the gamma-secretase quartet conducts Alzheimer's amyloid beta-peptide generation. *EMBO J.* 23, 483-488.

Ham, D., Skoryna, S.C. (2004). Cellular defense against oxidized low density lipoproteins and fibrillar amyloid beta in murine cells of monocyte origin with possible susceptibility to the oxidative stress induction. *Exp. Gerontol.* 39, 225-231.

Han, I-O., Kim, K-W., Ryu, J-H., Kim, W-K. (2002). p38 mitogen-activated protein kinase mediates lipopolysaccharide, not interferon-γ, -induced inducible nitric oxide synthase expression in mouse BV2 microglial cells. *Neurosci. Lett.* 325, 9-12.

Hanisch, U.K. (2002). Microglia as a source and target of cytokines. *Glia* 40, 140-155.

Hardy, J. (1997). Amyloid, the presenilins and Alzheimer's disease. *Trends. Neurosci.* 20, 154-159.

Hardy, J., Selkoe, D.J. (2002). The amyloid hypothesis of Alzheimer's disease: progress and problems on the road to therapeutics. *Science* 297, 353-356.

Harkany, T., Abraham, I., Konya, C., Nyakas, C., Zarandi, M., Penke, B., Lutten, P.G. (2000). Mechanism of beta-amyloid neurotoxicity: perspectives of pharmacotherapy. *Rev. Neurosci.* 11, 329-382.

Harmon, D. (1981). The aging process. *Proc. Natl. Acad. Sci. USA* 78, 7124-7128.

Harry, G.J., Lefebvre d'Hellencourt, C., Bruccoleri, A, Schmechel, D. (2000). Age-dependent cytokine responses: trimethyltin hippocampal injury in wild-type, APOE knockout, and APOE4 mice. *Brain Behav Immun.* 14, 288-304.

Hartley, D.M., Walsh, D.M., Ye, C.P., Diehl, T., Vasquez, S., Vassilev, P.M., Teplow, D.B., Selkoe, D.J. (1999). Protofibrillar intermediates of amyloid beta-protein induce acute electrophysiological changes and progressive neurotoxicity in cortical neurons. *J. Neurosci.* 19, 8876-8884.

Hartmann, T. (2001). Cholesterol, A beta and Alzheimer's disease. *Trends Neurosci.* 24, S45-S48.

Hashioka, S., Monji, A., Ueda, T., Kanba, S., Nakanishi, H. (2005). Amyloid-β fibril formation is not necessarily required for microglial activation by the peptides. *Neurochem. Int.* 47, 369-376.

Hauss-Wegrzyniak, B., Lukovic, L., Bigaud, M., Stoeckel, M.E. (1998). Brain inflammatory response induced by intracerebroventricular infusion of lipopolysaccharide: an immunohistochemical study. *Brain Res.* 794, 211-224.

Hayashida, T., Decaestecker, M., Schnaper, W. (2004). Cross-talk between ERK MAP kinase and Smad signaling pathways TGF-β dependent response in human mesangial cells. *FASEB J.* 17, 1576-1578.

Hemmer, K., Fransen, L., Vanderstichele, H., Vanmechelen, E., Heuschling, P. (2001). An in vitro model for the study of microglia-induced neurodegeneration: involvement of nitric oxide and tumor necrosis factor-α. *Neurochem. Int.* 38, 557-565.

Henrich-Noack, P., Prehn, J.H.M., Krieglstein, J. (1996). TGF-β1 protects hippocampal neurons against degeneration caused by transient global ischemia. Stroke 27, 1609-1614.

Herber, D.L., Roth, L.M., Wilson, N., Mason, J.E., Morgan, D., Gordon, M.N. (2004). Time-dependent reduction in Abeta levels after intracranial LPS administration in APP transgenic mice. *Exp. Neurol.* 190, 245-253.

Herber DL, Maloney JL, Roth LM, Freeman MJ, Morgan D, Gordon MN. (2005). Diverse microglial responses after intrahippocampal administration of lipopolysaccharide. *Glia*, Nov 15; [Epub ahead of print]

Herrera-Molina, R., von Bernhardi, R. (2005). Transforming growth factor-α1 produced by hippocampal cells modulates glial reactivity in culture. *Neurobiol Disease* 19, 229-236.

Herx, L.M., Rivest, S., Yong, V.W. (2000). Central nervous system-initiated inflammation and neurotrophism in trauma: IL-1β is required for the production of ciliary neurotrophic factor. *J. Immunol.* 165, 2232-1239.

Ho, L., Purohit, D., Haroutinian, V., Luterman, J.D., Willis, F., Naslund, J., Buxbaum, J.D., Mohs, R.C., Aisen, P.S., Pasinetti, G.M. (2001). Neuronal cyclooxigenase 2 expression in the hippocampal formation as a function of the clinical progression of Alzheimer disease. *Arch. Neurol.* 58, 487-492.

Hock, C., Heese, K., Hulette, C., Rosenberg, C., Otten, U. (2000). Region-specific neurotrophin imbalances in Alzheimer disease: decreased levels of brain-derived neurotrophic factor and increased levels of nerve growth factor in hippocampus and cortical areas. *Arch. Neurol.* 57, 846-851.

Hock, C., Konietzko, U., Streffer, J.R., Tracy, J., Signorell, A., Muller-Tillmanns, B., Lemke, U., Henke, K., Moritz, E., Garcia, E., Wollmer, M.A.Umbricht, D., de Quervain, D.J., Hofmann, M., Maddalena, A., Papassotiropoulos, A., Nitsch, R.M. (2003). Antibodies against beta-amyloid slow cognitive decline in Alzheimer's disease. *Neuron* 38, 547-554.

Hoek, R.M., Ruuls, S.R., Murphy, C.A., Wright, G.J., Goddard, R., Zurawski, S.M., Bloom, B., Homola, M.E., Streit, W.J., Brown, M.H., Barclay, A.N., Sedgwick, J.D. (2000). Down-regulation of the macrophage lineage through interaction with OX2 (CD2000). *Science* 290, 1768-1771.

Hofman, F.M., von Hanwher, R., Dinarello, C., Mizel, S., Hinton, D., Merrill, J.E. (1986). Immunoregulatory molecules and IL-2 receptors identified in multiple sclerosis brain. *J. Immunol.* 136, 3239-3245.

Hofmann, M.A., Drury, S., Fu, C., Qu, W., Taguchi, A., Lu, YT., Avila, C., Kambham, N., Bierhaus, A., Nawroth, P., Neurath, M.F., Slattery, T., Beach, D., McClary, J., Nagashima, M., Morser, J., Stern, D., Schmidt, A.M. (1999). RAGE mediates a novel proinflammatory axis: a central cell surface receptor for S100/calgranulin polypeptides. *Cell* 97, 889-901.

Hsu, H-Y., Chiu, S-L., Wen, M-H., Chen, K-Y., and Hua, K-F. (2001). Ligands of macrophage scavenger receptor induce cytokine expression via differential modulation of protein kinase signaling pathways. *J. Biol. Chem.* 276, 28719-28730.

Hu, J., Ferreira, A., Van Eldik, L.J. (1997). S100 beta induces neuronal cell death through nitric oxide release from astrocytes, *J. Neurochem.* 69, 2294-2301.

Hu, J., Akama, K., Krafft, G., Chromy, B., Van Eldik, L. (1998). Amyloid-β peptide activates cultured astrocytes: morphological alterations, cytokine induction and nitric oxide release. *Brain Res.* 785, 195-206.

Hu, J., Van Eldik, L.J. (1999). Glial-derived proteins activate cultured astrocytes and enhance beta amyloid-induced glial activation. *Brain Res.* 842, 46-54.

Husemann, J., Silverstein, S.C. (2001). Expression of scavenger receptor class B, type I, by astrocytes and vascular smooth muscle cells in normal adult mouse and human brain and in Alzheimer's disease brain. *Am. J. Pathol.* 158, 825-832

Husemann, J., Loike, J.D., Kodama, T., Silverstein, S.C. (2001). Scavenger receptor class B type I (SR-BI) mediates adhesion of neonatal murine microglia to fibrillar beta-amyloid. *J. Neuroimmunol.* 114, 142-150.

Husemann, J., Loike, J.D., Anankov, R., Febbraio, M., Silverstein, S.C. (2002). Scavenger receptors in neurobiology and neuropathology: their role on microglia and other cells of the nervous system. *Glia* 40, 195-205.

Ishida, Y., Kondo, T., Takayasu, T., Iwakura, Y., Mukaida, N. (2004). The essential involvement of cross-talk between IFN-γ and TGF-β1 in the skin wound-healing process. *J. Immunol.* 172, 1848-1855.

Ishii, K., Muelhauser, F., Liebl, U., Picard, M., Kuhl, S., Penke, B., Bayer, T., Weissler, M., Hennerici, M., Beyreuther, K., Hartmann, T., Fassbender, K. (2000). Subacute NO generation induced by Alzheimer's beta-amyloid in the living brain: reversal by inhibition of the inducible NO synthase, *FASEB J.* 14, 1485-1489.

Itzhaki, R.F., Wozniak, M.A., Appelt, D.M., Balin, B.J. (2004). Infiltration of the brain by pathogens causes Alzheimer's disease. *Neurobiol. Aging* 25, 619-627.

Jantzen, P.T., Connor, K.E., DiCarlo, G., Wenk, G.L., Wallace, J.L., Rojiani, A.M., Coppola, D., Morgan, D., Gordon, M.N. (2002). Microglial activation and beta-amyloid deposit reduction caused b a nitric oxide-releasing nonsteroidal anti-inflammatory drug in amyloid precursor protein plus presenilin-1 transgenic mice. J. Neurosci. 22, 2246-2254.

Jellinger, K.A. (1998). The neuropathological diagnosis of Alzheimer's disease. *J. Neural Transm. Suppl.* 53, 97-118.

Jick, H., Zornberg, G.L., Jick, S.S., Seshadri, S., Drachman, D.A. (2000). Statins and the risk of dementia. *Lancet* 356, 1627-1631.

Johnstone, M., Gearing, A.J.H., Miller, K.M. (1999). A central role for astrocytes in the inflammatory response to β-amyloid; chemokines, cytokines and reactive oxygen species are produced. *J. Neuroimmunol.* 93, 182-193.

Kakio, A., Nishimoto, S., Yanagisawa, K., Kozutsumi, Y., Matsuzaki, K. (2002). Interactions of amyloid beta-protein with various gangliosides in raft-like membranes: importance of GM1 ganglioside-bound form as an endogenous seed for Alzheimer amyloid. *Biochemistry* 41, 7385-7390.

Kakio, A., Nishimoto, S., Kozutsumi, Y., Matsuzaki, K. (2003). Formation of a membrane-active form of amyloid beta-protein in raft-like model membranes. *Biochem. Biophys. Res. Commun.* 303, 514-518.

Kalaria, R.N., Batí, S.U., Lust, W.D., Perry, G. (1993). The amyloid precursor protein in ischemic brain injury and chronic hypoperfusion. *Ann. N.Y. Acad. Sci.* 695, 190-193.

Kamal, A., Almenar-Queralt, A., LeBlanc, J.F., Roberts, E.A., Goldstein, L.S. (2001). Kinesis-mediated axonal transport of a membrane compartment containing beta-secretase and presenilin-1 requires APP. *Nature* 414, 643-648.

Kamenetz, F., Tomita, T., Hsieh, H., Seabrook, G., Borchelt, D., Iwatsubo, T., Sisodia, S., Malinow, R. (2003). APP processing and synaptic function. *Neuron* 37, 925-937.

Kayed, R., Head, E, Thompson, J.L., McIntire, T.M., Milton, S.C., Cotman, C.W., Glabe, C.G. (2003). Common structure of soluble amyloid oligomers implies common mechanism of pathogenesis. *Science* 300, 486-489.

Kelijman, M. (1991). Age-related alterations of the growth hormone/insulin-like growth factor-1 axis. J. Am. Geriatr. Soc. 39, 295-307.

Kiefer, R., Kreutzberger, G.W. (1990). Gamma interferon-like immunoreactivity in the rat nervous system. *Neurosci.* 37, 725-734.

Kim, W.S., Ordija, C.M., Freeman, M.W. (2003). Activation of signaling pathways by putative scavenger receptor class A (SR-A) ligands requires CD14 but not SR-A. *Biochem. Biophys. Res. Commun.* 310, 542-549

Kirchhoff, F., Dringen, R., Giaume, C. (2001). Pathways of neuron-astrocyte interactions and their possible role in neuroprotection. *Eur. Arch. Psychiatry Clin. Neurosci.* 251, 158-169.

Knopman, D.S., Parisi, J.E., Salviati, A., Floriach-Robert, M., Boeve, B.F., Ivnik, R.J., Smith, G.E., Dickson, D.W., Johnson, K.A., Petersen, L.E., McDonald, W.C., Braak, H., Petersen, R.C. (2003). Neuropathology of cognitively normal elderly. *J. Neuropathol. Exp. Neurol.* 62, 1087-1095.

Koudinov, A.R., Berezov, T.T. (2004). Alzheimer's amyloid-beta (A) is an essential synaptic protein, not neurotoxic junk. *Acta Neurobiol. Exp.* 64, 71-79.

Kovarik, P., Stoiber, D., Novy, M., Decker, T. (1998). Stat1 combines signals derived from IFN-gamma and LPS receptors during macrophage activation, EMBO J. 17, 3660-3668.

Kremer, J.J., Pallitto, M.M., Sklansky, D.J., Murphy, R.M. (2000). Correlation of beta-amyloid aggregate size and hydrophobicity with decreased bilayer fluidity of model membranes. *Biochemistry* 39, 10309-10318.

Krieger, M., Herz, J. (1994). Structures and functions of multiligand lipoprotein receptors: macrophage scavenger receptors and LDL receptor-related protein (LRP). *Ann. Rev. Biochem.* 63, 601-637

Krupinski, J., Kumar, P., Kumar, S., Kaluza, J. (1996). Increased expression of TGF- 1 in brain tissue after ischemic stroke in humans. *Stroke* 27, 852-857.

Kukull, W.A., Bowen, J.D. (2002). Dementia epidemiology. *Med. Clin. North Am.* 86, 573-590.

Lambert, M.P., Barlow, A.K., Chromy, B.A., Edwards, C., Freeds, R., Liosatos, M., Morgan, T.E., Rozovsky, I., Trommer, B., Viola, K.L. Wals, P., Zhang, C., Finch, C.E., Krafft, G.A., Klein, W.L. (1998). Diffusible, nonfibrillar ligands derived from Aβ1-42 are potent central nervous system neurotoxins. *Proc. Natl. Acad. Sci. USA* 95, 6448-6453.

Lander, H.M. (1997). An essential role for free radicals and derived species in signal transduction. *FASEB J.* 11, 118-124.

Lazarov-Spiegler, O., Solomon, A.S., Scwartz, M. (1998). Peripheral nerve-stimulated macrophages simulate a peripheral nerve-like regenerative response in rat transected optic nerve. *Glia.* 24, 329-337.

Ledeboer, A., Brevé, J.J.P., Poole, S., Tilders, F.J.H., Van Dam, A.M. (2000). Interleukin-10, interleukin-4, and transforming growth factor-β differentially regulate lipopolysaccharide-induced production of pro-inflammatory cytokines and nitric oxide in co-cultures of rat astroglial and microglial cells. *Glia* 30, 134-142.

Lee, S.C., Liu, W., Brosnan, C.F., Dickson, D.W. (1994). GM-CSF promotes proliferation of human fetal and adult microglia in primary cultures. Glia 12, 309-18.

Lee, H-G., Casadesus, G., Zhu, X., Takeda, A., Perry, G., Smith, M.A. (2004). Challenging the amyloid cascade hypothesis: senile plaques and amyloid-β as protective adaptations to Alzheimer disease. *Ann. N.Y. Acad. Sci.* 1019, 1-4.

Li, J.Y., Plomann, M., Brundin, P. (2003). Huntington's disease: a synaptopathy?. *Trends. Mol. Med.* 9, 414-420.

Lieb, K., Engels, S., Fiebich, B.L. (2003). Inhibition of LPS-induced iNOS and NO synthesis in primary rat microglial cells. *Neurochem. Int.* 42, 131-137.

Lim, G.P., Yang, F., Chu, T., Chen, P., Beech, W., Teter, B., Tran, T., Ubeda, O., Hsiao Ashe, K., Frautschy, S.A., Cole, G.M. (2000). Ibuprofen suppresses plaque pathology and inflammation in a mouse model of Alzheimer's disease. *J. Neurosci.* 20, 5709-5714.

Ling, E. Wong, W. (1993). The origin and nature of ramified and amoeboid microglia: A historical review and current concepts. *Glia* 7, 9–18.

Liu, Y., Dargusch, R., Schubert, D. (1997). Beta amyloid toxicity does not require RAGE protein. *Biochem. Biophys. Res. Commun.* 237, 37-40.

Liu, D., Smith, C.L., Barone, F.C., Ellison, J.A., Lyski, P.G., Li, K., Simpson, I.A. (1999). Astrocytes dismiss precedes delayed neuronal death in focal ischemic rat brain. *Brain Res. Mol. Brain Res.* 68, 29-41.

Liu, B., Du, L., Kong, L.Y., Hudson, P.M., Wilson, B.C., Chang, R.C., Abel, H.H., Hong, J. S. (2000). Reduction by naloxone of lipopolysaccharide-induced neurotoxicity in mouse cortical neuron-glia co-cultures. *Neurosci.* 97, 749-756.

Liu, B., Hong, J.S. (2003). Role of microglia in inflammation-mediated neurodegenerative diseases: mechanisms and strategies for therapeutic intervention. *J. Pharmacol. Exp. Ther.* 304, 1-7.

Loihl, A.K., Murphy, S. (1998). Expression of NOS-2 in glia associated with CNS pathology. *Prog. Brain Res.* 118, 253-267.

Lorenzo, A., Yankner, B.A. (1994). β-amyloid neurotoxicity requires fibril formation and is inhibited by Congo red. *Proc. Natl. Acad. Sci. USA* 91, 12243-12247.

Lorenzo, A., Yuan, M., Zhang, Z., Paganetti, P.A., Sturchler-Pierrat, C., Staufenbiel, M., Mautino, J., Vigo, F.S., Sommer, B., Yankner, B.A. (2000). Amyloid beta interacts with the amyloid precursor protein: a potential toxic mechanism in Alzheimer's disease. *Nat. Neurosci.* 3, 460-464.

Lovell, M.A., Gabbita, S.P., Markesbery, W.R. (1999). Increased DNA oxidation and decreased levels of repair products in Alzheimer's disease ventricular CSF. *J Neurochem.* 72, 771-776.

Lovell, M.A., Xie, C., Markesbery, W.R. (2001). Acrolein is increased in Alzheimer's disease brain and is toxic to primary hippocampal cultures. *Neurobiol. Aging* 22, 187-194.

Lue, L.F., Brachova, L., Walker, D.G., Shen, Y., Rogers, J. (1996). Characterization of glial cultures from rapid autopsies of Alzheimer's and control patients. *Neurobiol. Aging* 17, 421-429.

Lue, L.F., Rydel, R., Brigham, E.F., Yang, L.B., Hampel, H., Murphy, G.M. Jr., Brachova, L., Yan, S.D., Walker, D.G., Shen, Y., Rogers, J. (2001a). Inflammatory repertoire of Alzheimer's disease and nondemented elderly microglia in vitro. *Glia* 35, 72-79.

Lue, L.F., Walker, D.G., Brachova, L., Beach, T.G., Rogers, J., Schmidt, A.M., Stern, D.M., Yan, S.D. (2001b). Involvement of microglial receptor for advanced glycation endproducts (RAGE) in Alzheimer's disease: identification of a cellular activation mechanism. *Exp. Neurol.* 171, 29-45.

Luebke, J.I., Rosene, D.L. (2003). Aging alters dendritic morphology, input resistance, and inhibitory signaling in dentate granule cells of the rhesus monkey. *J. Comp. Neurol.* 460, 573-584.

Lukiw, W.J., Bazan, N.G. (2000). Neuroinflammatory signalling upregulation in Alzheimer's disease. *Neurochem. Res.* 25, 1173-1184.

Luterman, J.D., Haroutunian, V., Yemul, S., Ho, L., Purohit, D., Aisen, P.S., Mohs, R., Pasinetti, G.M. (2000). Cytokine gene expression as a function of the clinical progression of Alzheimer disease dementia. *Arch. Neurol.* 57, 1153-1160.

Luth, H., Munch, G., Arendt, T. (2002). Aberrant expression of NOS isoforms in Alzheimer's disease is structurally related to nitrotyrosine formation, *Brain Res.* 953, 135-143.

Malchiodi-Albedi, F., Domenici, M.R., Paradisi, S., Bernardo, A., Ajmone-Cat, M.A., Minghetti, L. (2001). Astrocytes contribute to neuronal impairment in Aβ toxicity increasing apoptosis in rat hippocampal neurons. *Glia* 34, 68-72.

Makranz, C., Cohen, G., Baron, A., Levidor, L., Kodama, T., Reichert, F. And Rotshenker, S. (2004). Phosphatidylinositol 3-kinase, phosphoinositide-specific phospholipase-Cgamma and protein kinase-C signal myelin phagocytosis mediated by complement receptor-3 alone and combined with scavenger receptor-AI/II in macrophages. *Neurobiol. Disease* 15, 279-286.

Marcus, J., Karackattu, S., Fleegal, M., Sumners, C. (2003). Cytokine-stimulated inducible nitric oxide synthase expression in astroglia: Role of Erk Mitogen-activated protein kinase and NF-κB. *Glia* 41, 152-160.

Markesbery, W.R., Lovell, M.A. (1998). Four-hydroxynonenal, a product of lipid peroxidation, is increased in the brain in Alzheimer's disease. *Neurobiol. Aging* 19, 33-36.

Martin, L., Pardo, C., Cork, L., Price, D. (1994). Synaptic pathology and glial responses to neuronal injury precede the formation of senile plaques and amyloid deposits in the aging cerebral cortex. *Am. J. Pathol.* 145, 1358-1381.

Marzolo, M.P., von Bernhardi, R., Inestrosa, N.C. (1999). Expression of alpha(2)-macroglobulin receptor/low density lipoprotein receptor-related protein (LRP) in rat microglial cells. *J. Neurosci. Res.* 58, 387-95.

Matsunaga, S., Nagano, S., Onishi, T., Morimoto, N., Suzuki, S., Komiya, S. (2003). Age-related changes in expression of transforming growth factor-β and receptors in cells of intervertebral discs. *J. Neurosurg.* 98, 63-67.

Matsuoka, Y., Picciano, M., Malester, B., LaFrancois, J., Zehr, C., Daeschner, J.M., Olschowka, J.A., Fonseca, M.I., O'Banion, M.K., Tenner, A.J., Lemere, C.A., Duff, K. (2001). Inflammatory responses to amyloidosis in a transgenic mouse model of Alzheimer's disease. *Am. J. Pathol.* 158, 1345-1354.

Matter, M.L., Zhang, Z., Nordstedt, C., Ruoslahti, E. (1998). The alpha5beta1 integrin mediates elimination of amyloid-beta peptide and protects against apoptosis. *J. Cell Biol.* 141, 1019-1030.

Mattson, M.P. (2000). Apoptosis in neurodegenerative disorders. *Mol. Cell Biol. Reviews* 1, 120-129.

McCann, S.M. (1997). The nitric oxide hypothesis of brain aging. *Exp. Geront.* 32(4-5), 431-440.

McCartney-Francis, N.L., Wahl, S.M. (2002). Dysregulation of IFN-γ signaling pathways in the absence of TGF-β1. *J Immunol.* 169, 5941-5947.

McDonalds, D.R., Bamberger, M.E., Combs, C.K., Landreth, G.E. (1998). beta-Amyloid fibrils activate parallel mitogen-activated protein kinase pathways in microglia and THP1 monocytes, *J Neurosci.* 18, 4451-4460.

McGeer, P.L., McGeer, E.D. (1995). The inflammatory response system of brain: implications for therapy of Alzheimer and other neurodegenerative diseases. *Brain. Res. Rev.* 21, 195-218.

McGeer, P.L., McGeer, E.G. (2001). Inflammation, autotoxicity and Alzheimer's disease. *Neurobiol. Aging* 22, 799-809.

McMillian, M.K., Thai, L., Hong, J.S., O'Callaghan, J.P., Pennypacker, K.R. (1994). Brain injury in a dish: a model for reactive gliosis. *TINS* 17, 138-142.

Meda, L., Cassatella, M.A., Szendrei, G.I., Otvos, J.L., Baron, P., Villalba, M., Ferrari, P., Rossi, F. (1995). Activation of microglial cells by beta-amyloid protein and interferon-gamma. *Nature*. 374, 647-650.

Meda, L., Bernasconi, S., Bonaiuto, C., Sozzani, S., Zhou, D., Otvos, J.L., Mantovani, A., Rossi, F., Cassatella, M.A. (1996). β-amyloid[25-35] peptide and INFγ synergistically induce the production of the chemotactic cytokine MCP-1/JE in monocytes and microglial cells, *J. Immunol.* 157, 1213-1218.

Meda, L., Baron, P., Prat, E., Scarpini, E., Scarlato, G., Cassatella, M.A., Rossi, F. (1999). Proinflammatory profile of cytokine production by human monocytes and murine microglia stimulated with beta-amyloid[25-35], *J Neuroimmunol.* 93, 45-52.

Meda, L., Baron, P., Scarlato, G. (2001). Glial activation in Alzheimer's disease: the role of Aβ and its associated proteins. *Neurobiol. Aging.* 22, 885-893.

Mehlhorn, G., Hollbron, M., Schliebs, R. (2000). Induction of cytokines in glial cells surrounding cortical beta-amyloid plaques in transgenic Tg2576 mice with Alzheimer pathology. *Int. Dev. Neurosci.* 18, 423-431.

Melton, L.M., Keith, A.B., Davis, S., Oakley, A.E., Edwardson, J.A., Morris, C.M. (2003). Chronic glial activation, neurodegeneration, and APP immunoreactive deposits following acute administration of double-stranded RNA. *Glia* 44, 1-12.

Michikawa, M. (2003). The role of cholesterol in pathogenesis of Alzheimer's disease: dual metabolic interaction between amyloid beta-protein and cholesterol. *Mol. Neurobiol.* 27, 1-12.

Mittaud, P., Labourdette, G., Zingg, H., Guenot-Di Scala, D. (2002). Neurons modulate oxytocin receptor expression in rat cultured astrocytes: Involvement of TGF-β and membrane components. *Glia* 37, 169-177.

Miyata, M., Smith, J.D. (1996). Apolipoprotein E allele-specific antioxidant activity and effects on cytotoxicity by oxidative insults and beta-amyloid peptides. *Nat. Genet.* 14, 55-61.

Mouri, H., Sakaguchi, K., Sawayama, T., Senoh, T., Ohta, T., Nishimura, M., Fujiwara, A., Terao, M., Shiratori, Y., Tsuji, T. (2002). Suppressive effects of transforming growth factor-β1 produced by hepatocellular carcinoma cell lines on interferon-γ production by peripheral blood mononuclear cells. *Acta Med. Okayama* 56, 309-315.

Mrak, R., Sheng, J., Griffin, S. (1995). Glial cytokines in Alzheimer's disease. *Human Pathology* 26, 816-823.

Mrak, R., Griffin, S. (2001). The role of activated astrocytes and of the neurotrophic cytokine S100B in the pathogenesis of Alzheimer's disease. *Neurobiol. Aging* 22, 915-922.

Murray, C.A., McGahon, B., McBennet, S., Lynch, M. (1997). Interleukin-1 beta inhibits glutamate release in hippocampus of young, but not aged, rats. *Neurobiol. Aging* 18, 343-348.

Murray, C.A. Lynch, M. (1998a). Evidence that increased hippocampal expression of the cytokine interleukin-1 beta is a common trigger for age- and stress-induced impairments in long-term potentiation. *J. Neurosci.* 18, 2974-2981.

Murray, C.A. Lynch, M. (1998b). Dietary supplementation with vitamin E reverses the age-related deficit in long term potentiation in dentate gyrus. *J. Biol. Chem.* 273, 12161-12168.

Nagele, R.G., Wegiel, J., Venkatamaran, V., Imaki, H., Wang, K.C., Wegiel. (2004). Contribution of glial cells to the development of amyloid plaques in Alzheimer's disease. *Neurobiol. Aging* 25, 663-674.

Naslund, J., Haroutunian, V., Mohs, R., Davis, K.L., Davies, P., Greengard, P., Buxbaum, J.D. (2000). Correlation between elevated levels of amyloid beta-peptide in the brain and cognitive decline. *JAMA* 283, 1571-1577.

Nedergaard, M. (1994). Direct signaling from astrocytes to neurons in cultures of mammalian brain cells. *Science* 263, 1768-1771.

Neumann, H., Schmidt, H., Wilharm, E., Behrens, L., Wekerle, H. (1997). Interferon γ gene expression in sensory neurons: evidence for autocrine gene regulation. *J. Exp. Med.* 186, 2023-2031.

Neumann, H., Wekerle, H. (1998). Neuronal control of the immune response in the central nervous system: Linking brain immunity to neurodegeneration. *J. Neuropathol. Exp.* Neurol. 57, 1-9.

Neuroinflammation Working group. (2000). Inflammation and Alzheimer's disease. *Neurobiol. Aging* 21, 383-421.

Nguyen, M.D., Julien, J.-P., Rivest, S. (2002). Innate immunity: the missing link in neuroprotection and neurodegeneration? *Nature Rev. Neurosci.* 3, 216-227.

Nichols, N.R. (1999). Glial responses to steroids as markers of brain aging. *J. Neurobiol.* 40, 585-601.

Nicoll, J.A., Wilkinson, D., Holmes, C., Steart, P., Markham, H., Weller, R.O. (2003). Neuropathology of human Alzheimer disease after immunization with amyloid-beta peptide: a case report. *Nat. Med.* 9, 448-452.

Nunomura, A., Perry, G., Aliev, G., Iría, K., Takeda, A., Balraj, E.K., Jones, P.K., Ghanbari, H., Wataya, T., Shimohama, S., Chiba, S., Atwood, C.S., Petersen, R.B., Smith, M.A. (2001). Oxidative damage is the earliest event in Alzheimer disease. *J. Neuropathol. Exp.* Neurol. 60, 759-767.

Obermayr, R.P., Mayerhofer, L., Knechtelsdorfer, M., Merisch, N., Huber, E.R., Geyer, G., Tragl, K-H. (2005). The age-related down-regulation of the growth hormone/insulin-like growth factor-1 axis in the elderly male is reversed considerably by donepezil, a drug for Alzheimer's disease. *Exp. Gerontol.* 40, 157-163.

O'Brien, M.F., Lenke, L.G., Lou, J., Bridwell, K.H., Joyce, M.E. (1994). Astrocyte response and transforming growth factor-beta localization in acute spinal cord injury. *Spine* 19, 2321-2329.

Olsson, T., Kelic, S., Edlund, C., Bakhiet, M., Höjeberg, B., van der Meide, P.H., Lungdahl, A., Kristensson, K. (1994). Neuronal interferon-γ immunoreactive molecule: bioactivities and purification. *Eur. J. Immunol.* 24, 308-314.

Pahan, K., Sheikh, F.G., Namboodiri, Am., Singh, I. (1997). Lovastatin and phenylacetate inhibit the induction of nitric oxide synthase and cytokines in rat primary astrocytes, microglia, and macrophage. *J. Clin. Invest.* 100, 2671-2679.

Palop, J.J. Jones, B., Kekonius, L., Chin, J., Yu, G.Q., Raber, J., Masliah, E., Mucke, L. (2003). Neuronal depletion of calcium-dependent proteins in the dentate gyrus is tightly linked to Alzheimer's disease-related cognitive deficits. *Proc. Natl. Acad. Sci. USA* 100, 9572-9577.

Paresce, D.M., Ghosh, R.N., Maxfield, F.R. (1996). Microglial cells internalize aggregates of the Alzheimer's disease amyloid beta-protein via a scavenger receptor. *Neuron* 17, 553-565.

Paresce, D.M., Chung, H., Maxfield, F.R. (1997). Slow degradation of aggregates of the Alzheimer's disease amyloid β-protein by microglial cells. *J. Biol. Chem.* 272, 29390-29397.

Parsons, C.G., Danysz, W., Quack, G. (1999). Memantine is a clinically well tolerated N-methyl-D-aspartate (NMDA) receptor antagonist – a review of preclinical data. *Neuropharmacology* 38, 735-767.

Pasinetti, G.L. (1998). Cyclooxigenase and inflammation in Alzheimer's disease: Experimental approaches and clinical interventions. *J. Neurosci. Res.* 54, 1-6.

Pawate, S., Shen, Q., Fan, F., Bhat, N.R. (2004). Redox regulation of glial inflammatory response to lipopolysaccharide and interferonγ. *J. Neurosci. Res.* 77, 540-551.

Penkowa, M., Girallt, M., Carrasco, J., Hanberd, H., Hidalgo, J. (2000). Impaired inflammatory stress and neurodegeneration after brain injury in interleukin-6-deficient mice. *Glia* 32, 271-285.

Pike, C.J., Walencewicz, A.J., Glabe, C.G., Cotan, C.W. (1991). In vitro aging of β-amyloid protein causes peptide aggregation and neurotoxicity. *Brain Res.* 563, 311-314.

Plant, L.D., Boyle, J.P., Smith, I.F., Peers, C., Pearson, H.A. (2003). The production of amyloid beta peptide is a critical requirement for the viability of central neurons. *J. Neurosci.* 23, 5531-5535.

Prehn, J.H.M., Bindokas, W.P., Jordán, J., Galindo, M.F., Ghadge, G.D., Roos, R.P., Boise, L.H., Thompson, C.B., Krajewski, S., Reed, J.C., Miller, R.J. (1996). Protective effect of transforming growth factor-β1 on β-amyloid neurotoxicity in rat hippocampal neurons. *Molec. Pharmacol.* 49, 319-328.

Prinz, M., Kann, O., Draheim, H.J., Schumann, R.R., Kettenmann, H., Webwer, J.R., Hanisch, U.K. (1999). Microglial activation by components of gram-positive and – negative bacteria: distinct and common routes to the induction of ion channels and cytokines. *J. Neuropathol. Exp. Neurol.* 58, 1078-1089.

Pyo, H., Jou, I., Jung, S., Hong, S., Joe, E-H. (1998). Mitogen-activated protein kinases activated by lipopolysaccharide and β-amyloid in cultured rat microglia. *NeuroReport* 9, 871-874.

Qiu, Z., Sttickland, D.K., Hyman, B.T., Rebeck, G.W. (2001). Elevation of LDL receptor-related protein levels via ligand interactions in Alzheimer's disease and in vitro. J. *Neuropathol. Exp. Neurol.* 60, 430-440.

Rada, P., Mark, G.P., Vitek, M.P., Manago, R.M., Blume, A.J., Beer, B., Hoebel, B.G. (1991). Interleukin-1 beta decreases acetylcholine measured by microdialysis in the hippocampus of freely moving rats. *Brain Res.* 550, 287-290.

Ramírez, G., Toro, R., Döbeli, H., von Bernhardi, R. (2005). Protection of rat primary hippocampal cultures from Aß cytotoxicity by pro-inflammatory molecules is mediated by astrocytes. *Neurobiol Disease* 19, 243-254.

Re, F., Belyanskaya, S.L., Riese, R.J., Cipriani, B., Fischer, F.R., Granucci, F., Ricciardi-Castagnoli, P., Brosnam, C., Stern, L.J., Strominger, J.L., Santambrogio, L. (2002). Granulocyte-macrophage colony-stimulating factor induces an expression program in neonatal microglia that primes them for antigen presentation. *J. Immunol.* 169, 2264-2273.

Reale, M., Iarlori, C., Gambi, F., Feliciani, C., Salone, A., Toma, L., DeLuca, G., Salvatore, M., Conti, P., Gambi, D. (2004). Treatment with an acetylcholinesterase inhibitor in Alzheimer' patients modulates the expression and production of the pro-inflammatory and anti-inflammatory cytokines. *J. Neuroimmunol.* 148, 162-171.

Refolo, L.M., Malester, B., LaFrancois, J., Bryant-Thomas, T., Wang, R., Tint, G.S., Sambamurti, K., Duff, K., Pappolla, M.A. (2000). Hypercholesterolemia accelerates the Alzheimer's amyloid pathology in a transgenic mouse model. *Neurobiol. Dis.* 7, 321-331.

Reichert, F., Rotshenker, S. (1996). Deficient activation of microglia during optic nerve degeneration. *J. Neuroimmunol.* 70, 153-161.

Reisberg, B., Doody, R., Stoffler, A., Schmitt, F., Ferris, S., Mobius, H.J. (2003). Memantine in moderate-to-severe Alzheimer's disease. *N. Engl. J. Med.* 348, 1333-1341.

Reynolds, J.L., Ignatowski, T.A., Sud, R., Spengler, R.N. (2004). Brain-derived tumor necrosis factor-alpha and its involvement in noradrenergic neuron functioning involved in the mechanism of action of an antidepressant. *J. Pharmacol. Exp. Ther.* 310, 1216-1225.

Rockwood, K., Kirkland, S., Hogan, D.B. et al. (2002). Use of lipid-lowering agents, indication bias, and the risk of dementia in community-dwelling elderly people. *Arch. Neurol.* 59, 223-227.

Rogers, J.T., Leiter, L.M., McPhee, J., Cahill, C.M., Zhan, S.S., Potter, H., Nilsson, L.N. (1999). Translation of the Alzheimer amyloid precursor protein mRNA is up-regulated by interleukin-1 through 5'-untranslated region sequences. *J. Biol. Chem.* 274, 6421-6431.

Rogers, J., Lue, L.F. (2001). Microglial chemotaxis, activation, and phagocytosis of amyloid beta-peptide as linked phenomena in Alzheimer's disease. *Neurochem. Int.* 39, 333-340.

Rogers, J., Strohmeyer, R., Kovelowski, C.J., Li, R. (2002). Microglia and inflammatory mechanisms in the clearance of amyloid β peptide. *Glia* 40, 260-269.

Roses, A.D. (1996). Apolipoprotein E in neurology. *Curr. Opin. Neurol.* 4, 265-270.

Ross, F.M., Allan, S.M., Rothwell, N.J., Verkhratsky, A. (2003). A dual role for interleukin-1 in LTP in mouse hippocampal slices. *J. Neuroimmunol.* 144, 61-67.

Roth, A.D., Ramírez, G., Alarcón, R., von Bernhadi, R. (2005). Oligodendrocytes damage in Alzheimer's disease: Beta amyloid toxicity and inflammation. *Biol. Res.* 38, 381-387.

Rothwell, N., Luheshi, G. (2000). Interleukin 1 in the brain: Biology, pathology and therapeutic target. *Trends in Neurosci.* 23, 618-625.

Rubin, L.L. (1998). Neuronal Cell Death: An updated view. In: Van Leeuwen F.W., Salehi, A., Giger, R.J., Holtmaat, A.J.G.D., Verhagen, J. (Eds.), *Progress in Brain Research*, Elsevier, Amsterdam, vol 117, pp. 3-8.

Sairanen, T., Ristimaki, A., Karjaleinen-Lindsberg, M.L. Paetau, A., Kaste, M., Lindsberg, P.J. (1998). Cyclooxigenase-2 is induced globally in infracted human brain. *Ann. Neurol.* 43, 738-747.

Sala, G., Galimberti, G., Canevari, C., Raggi, M.E., Isella, V., Facheris, M., Appollonio, I., Ferrarese, C. (2003). Peripheral cytokine release in Alzheimer patients: correlation with disease severity. *Neurobiol. Aging* 24, 909-914.

Salina, S., Díaz, R., Abraham, N.G., Ruiz de Galarreta, C.M., Cuadrado, A. (2003). Nerve growth factor protects against 6-hydroxydopamine-induced oxidative stress by increasing expression of heme oxigenase-1 in a phosphatidylinositol 3-kinase-dependent manner. *J. Biol. Chem.* 278, 13898-13904.

Sasaki, A., Yamaguchi, H., Ogawa, A., Sugihara, S., Nakasato, Y. (1997). Microglial activation in early stages of amyloid beta protein deposition. *Acta Neuropathol.* (Berl.) 94, 316-322.

Sastre, M., Dewachter, I., Landreth, G.E., Wilson, T.M., Klockgether, T., Van Leuven, F., Heneka, M.T. (2003). Nonsteroidal anti-inflammatory drugs and peroxisome proliferator-activated receptor-gamma agonists modulate immunostimulated processing of amyloid precursor protein through regulation of beta-secretase. *J. Neurosci.* 23, 9796-9804.

Saud, K., Herrera-Molina, R., von Bernhardi, R. (2005). Pro- and anti-inflammatory cytokines regulate the ERK pathway: implication of the timing for the activation of microglial cells. *Neurotox. Res.* 8, 277-287.

Scheff, S.W., Price, D.A. (2003). Synaptic pathology in Alzheimer's disease: a review of ultrastructural studies. *Neurobiol. Aging.* 24, 1029-1046.

Schenk, D., Barbour, R., Dunn, W., Gordon, G., Grajeda, H., Guido, T., Hu, K., Huang, J., Johnson-Wood, K., Khan, K., Kholodenko, D., Lee, M., Liao, Z., Lieberburg, I., Motter, R., Mutter, L., Soriano, F., Shopp, G., Vasquez, N., Vandevert, C., Walker, S., Wogulis, M., Yednock, T., Games, D., Seubert, P. (1999). Immunization with amyloid-beta attenuates Alzheimer-disease-like pathology in the PDAPP mouse. *Nature* 400, 173-177.

Schmidt, B., Stoll, G., Toyka, K.V., Hartung, H.-P. (1990). Rat astrocytes express interferon-γ immunoreactivity in normal optic nerve and after nerve transection. *Brain Res.* 515, 347-350.

Schubert, P., Morino, T., Miyazaki, H., Ogata, T., Nakamura, Y., Marchini, C., Ferroni, S. (2000). Cascading glia reactions: A common pathomechanism and its differentiated control by cyclic nucleotide signalling. *Ann. N. Y. Acad. Sci.* 903, 24-33.

Selkoe, D.J. (1996). Amyloid beta-protein and the genetics of Alzheimer's disease. *J. Biol. Chem.* 271, 18295-18298.

Selkoe, D.J. (2000). The origins of Alzheimer's disease: a is for amyloid. *JAMA.* 283, 1615-1617.

Selkoe, D. (2001). Alzheimer's disease: genes, proteins and therapy. *Physiol. Rev.* 81, 741-766.

Selkoe, D.J. (2002). Alzheimer's disease: is a synaptic failure. *Science* 298, 789-791.

Seubert, P., Vigo-Pelfrey, C., Esch, F., Lee, M., Dovey, H., Davis, D., Sinha, S., Schlossmacher, M., Whaley, J., Swindlehurst, C., McCormack, R., Wolfert, P., Selkoe, D.J., Lieberburg, I., Schenk, D. (1992). Isolation and quantification of soluble Alzheimer's beta-peptide from biological fluids. *Nature* 359, 325-327.

Shaffer, L.M., Dority, M.D., Grupta-Bansal, R., Frederickson, R.C., Youkin, S.G., Brunden, K.R. (1995). Amyloid beta protein (Abeta) removal by neurological cells in culture. *Neurobiol. Aging* 16, 737-745.

Sheng, J.G., Mrak, R.E., Griffin, W.S. (1995). Microglial interleukin-1 alpha expression in brain regions in Alzheimer's disease: correlation with neuritic plaque distribution. *Neuropathol. Appl. Neurobiol.* 21, 290-301.

Sheng, J.G., Mrak, R.E., Griffin, W.S. (1997). Neuritic plaque evolution in Alzheimer's disease is accompanied by transition of activated microglia from primed to enlarged to phagocytic forms. *Acta Neuropathol.* (Berl) 94, 1-5.

Sheng, J.G., Bora, S.H., Xu, G., Borchelt, D.R., Price, D.L., Koliatos, V.E. (2003). Lipopolysaccharide-induced-neuroinflammation increases intracellular accumulation of amyloid precursor protein and amyloid beta peptide in APPswe transgenic mice. *Neurobiol. Dis.* 14, 133-145.

Simons, M., Keller, P., De Strooper, B., Beyreuther, K., Dotti, C.G., Simmons, K. (1998). Cholesterol depletion inhibits the generation of beta-amyloid in hippocampal neurons. *Proc. Natl. Acad. Sci. USA.* 95, 6460-6464.

Smith, M.A., Hirai, K., Hsiao, K., Pappolla, M.A., Harris, P.L., Siedlak, S.L., Tabaton M, Perry G. (1998). Amyloid-beta deposition in Alzheimer transgenic mice is associated with oxidative stress. *J. Neurochem.* 70, 2212-2215.

Smits, H.A., Boven, L.A., Pereira, C.F., Verhoef, J., Nottet, H.S. (2000). Role of macrophage activation in the pathogenesis of Alzheimer's disease and human immunodeficiency virus type 1-associated dementia, *Eur. J. Clin. Invest.* 30, 526-535.

Smits, H.A., van Beelen, A.J., de Vos, N.M., Rijmus, A., van der Bruggen, T., Verhoef, J., van Muiswinkel, F.L., Nottet, H.S. (2001). Activation of human macrophages by amyloid-beta is attenuated by astrocytes. *J. Immunol.* 166, 6869-6876.

Smits, H.A., Rijsmus, A., van Loon, J.H., Wat, J.W., Verhoef, J., Boven LA, Nottet, H.S. (2002). Amyloid-beta-induced chemokine production in primary human macrophages and astrocytes, J. *Neuroimmunol.* 127, 160-168.

Sondag, C.M., Combs, C.K. (2004). Amyloid precursor protein mediated proinflammatory activation of monocytic lineage cells. *J. Biol. Chem.* 279, 14456-14463.

Sortino, M.A., Chisari, M., Merlo, S., Vancheri, C., Caruso, M., Nicoletti, F., Canonico, P.L., Copan, A. (2004). Glia mediates the neuroprotective action of stradiol on beta-amyloid-induced neuronal death. *Endocrinology* 145, 5080-5086.

Sperlagh, B., Baranyi, M., Hasko, G., Vizi, E.S. (2004). Potent effect of interleukin-1 beta to evoke ATP and adenosine release from rat hippocampal slices. *J. Neuroimmunol.* 151, 33-39.

Stephan, A., Laroche, S., Davis, S. (2001). Generation of aggregated β-amyloid in the rat hippocampus impairs synaptic transmission and plasticity and causes memory deficits. *J. Neurosci.* 21, 5703-5714.

Stewart, W.F., Kawas, C., Corrada, M., Metter, E.J. (1997). Risk of Alzheimer's disease and duration of NSAID use. *Neurology* 48, 626-632.

Streit, W.J. (2002). Microglia as neuroprotective, immunocompetent cells of the CNS. *Glia* 40, 133-139.

Streit, W.J., Sammons, N.W., Kuhns, A.J., Sparks, D.L. (2004). Dystrophic microglia in the aging human brain. *Glia* 45, 208-212.

Sudo, S., Tanaka, J., Toku, K., Desaki, J., Matsuda, S., Arai, T., Sakanaka, M., Maeda, N. (1998). Neurons induce the activation of microglial cell in vitro. *Exp. Neurol.* 154, 499-510.

Takahashi, M., Dore, S., Ferris, C.D., Tomita, T., Sawa, A., Wolosker, H., Borchelt, D.R., Iwatsubo, T., Kim, S.H., Thinakaran, G., Sisodia, S.S., Snyder, S.H. (2000). Amyloid precursor protein inhibit heme oxygenase activity and augment neurotoxicity in Alzheimer's disease. *Neuron* 28, 461-473.

Takahashi, R.H., Almeida, C.G., Kearney, P.F., Yu, F., Lin, M.T., Milner, T.A., Gouras, G.K. (2004). Oligomerization of Alzheimer's beta-amyloid within processes and synapses of cultured neurons and brain. *J. Neurosci.* 24, 3592-3599.

Talley, A., Dewhurst, S., Perry, S., Dollars, S., Gummuluru, S., Fine, S., New, D., Epstein, L., Gendelman, H., Gelbard, H. (1995). Tumor necrosis alpha-induced apoptosis in human neuronal cells: protection by the antioxidant N-acetylcysteine and the genes bcl-2 and crmA. *Mol. Cell. Biol.* 15, 2359-2366.

Tanaka, M., Sotumatsu, A., Yoshida, T., Hirai, S., Nishida, A. (1994). Detection of superoxide production by activated microglia using a sensitive and specific chemiluminescence's assay and microglia-mediated PC12 cell death. *J. Neurochem.* 63, 266-270.

Terzi, E., Holzemann, G., Seeling, J. (1997). Interaction of Alzheimer beta-amyloid peptide(1-40) with lipid membranes. *Biochemistry* 36, 14845-14852.

Tiffany, H.L., Lavigne, M.C., Cui, Y.H., Wang, J.M., Leto, T.L., Gao, J.L., Murphy, P.M. (2001). Amyloid-beta induces chemotaxis and oxidant stress by acting at formylpeptide receptor 2, a G protein-coupled receptor expressed in phagocytes and brain. *J. Biol. Chem.* 276, 23645-23652.

Tran, M.H., Yamada, K., Nabeshima, T. (2002). Amyloid beta-peptide induces chollinergic dysfunction and cognitive deficit: a minireview. *Peptides* 23, 1271-1283.

Traugott, U., Lebon, P. (1988). Interferon-γ and Ia antigen are present on astrocytes in active chronic multiple sclerosis lesions. *J. Neurol. Sci.* 84, 257-264.

Trejo, J.L., Carro, E., Garcia-Galloway, E., Torres-Aleman, I. (2004). Role of insulin-like growth factor-I signaling in neurodegenerative disease. J. Mol. Med. 82, 156-162.

Uylings, H.B., de Brabander, J.M. (2002). Neuronal changes in normal human aging and Alzheimer's disease. *Brain Cogn.* 49, 268-276.

Van Groen, T., Kadish, I. (2005). Transgenic AD model mice, effects of potential anti-AD treatments on inflammation and pathology. *Brain Res. Revs.* 48, 370-378.

Vaughan, C.J., Delanty, N. (1999). Neuroprotective properties of statins in cerebral ischemia and stroke. *Stroke* 30, 1969-1973.

Vegeto, E., Bonincontro, C., Polli, G., Sala, A, Viappiani, S., Nardi, F., Brusadelli, A., Viviani, B., Ciana, P., Maggi, A. (2001). Estrogen prevents the lipopolysaccharide-induced inflammatory response in microglia. *J. Neurosci.* 21, 1809-1818.

Vermeer, S.E., Prins, N.D., den Heijer, T.,Hofman, A., Koudstaat, P.J., Breteler, M.M. (2003). Silent brain infarcts and the risk of dementia and cognitive decline. *N. Engl. J. Med.* 348, 1215-1222.

Vincent, A.M., Maiese, K. (1999a). Nitric oxide induction of neuronal endonuclease activity in programmed cell death. *Exp. Cell Res.* 246, 290-300.

Vincent, A.M., Maiese, K. (1999b). Direct temporal analysis of apoptosis induction in living adherent neurons. J. Histochem. *Cytochem.* 47, 661-672.

Vincent, V.A., Tielders, F.J., Van Dam, A.M. (1997). Inhibition of endotoxin-induced nitric oxide synthase production in microglial cells by the presence of astroglial cells: a role for transforming growth factor β. *Glia* 19, 190-198.

Vincent, V.A., Lowik, C.W., Verheijen, J.H., de Bart, A.C., Tilders, F.J., Van Dam AM. (1998). Role of astrocyte-derived tissue-type plasminogen activator in the regulation of endotoxin-stimulated nitric oxide production by microglial cells. *Glia* 22, 130-137.

Vincent, V.A., Selwood, S.P., Murphy, G.M. Jr. (2002). Proinflammatory effects of M-CSF and A-beta in hippocampal organotypic cultures. *Neurobiol. Aging.* 23, 349-362.

Vitek, M.P., Bhattacharya, K., Glendening, J.M. Stopa, E., Vlassara, H., Bucala, R., Monogue, K., Cerami, A. (1994). Advanced glycation end products contribute to amyloidosis in Alzheimer's disease. *Proc. Natl. Acad. Sci. USA* 91, 4766-4770.

Vitkovic L., Bockaert, J., Jacque, C. (2000). "Inflammatory" Cytokines: Neuromodulators in normal Brain? *J. Neurochem.* 74, 457-471.

Vitolo, O.V., Sant'Angelo, A., Constanzo, V., Battaglia, F., Arancio, O., Shelanski, M. (2002). Amyloid beta-peptide inhibition of the PKA/CREB pathway and long-term potentiation: reversibility by drugs that enhance cAMP signaling. *Proc. Natl. Acad. Sci. USA* 99, 13217-13221.

von Bernhardi, R., Ramírez, G., Matile, H. Döbeli, H. (2001). Immobilized APP constructs: a tool for the in vitro screening of glial cells reactivity. *Eur J. Neurosci.* 14, 946-956.

von Bernhardi, R., Ramírez, G. (2001). Microglia-astrocyte interaction in Alzheimer's disease: friends or foes for the nervous system? *Biol Res.* 34, 123-128.

von Bernhardi, R., Ramírez, G., De Ferrari G.V. and Inestrosa, N.C. (2003). Acetylcholinesterase induces the expression of the β-Amyloid Precursor Protein in glia and activates glial cells in culture. *Neurobiol Disease*, 14(3), 447-57.

von Bernhardi, R., Eugenín, J. (2004). Microglia - astrocyte interaction in Alzheimer's disease: modulation of cell reactivity to Aβ. *Brain Res.* 1025, 186-193.

von Bernhardi R. (2005). Aging: Biochemistry and functional changes of the Central Nervous System. Rev. Chil. Neuro-Psiq. 43(4), 297-304.

Wallace, M.N., Geddes, J.G., Farquhar, D.A., Masson, M.R. (1997). Nitric oxide synthase in reactive astrocytes adjacent to beta-amyloid plaques. *Exp. Neurol.* 144, 266-272.

Walsh, D.M., Hartley, D.M., Kusumoto, Y., Fezoui, Y., Condrom, M.M., Lomakin. A., Benedeck, G.B., Selkoe, D.J. (1999). Amyloid beta-protein fibrillogenesis. Structure and biological activity of protofibrillar intermediates. *J. Biol. Chem.* 274, 25945-25952.

Walsh, D.M., Tseng, B.P., Rydel, R.E., Podisny, M.B., Selkoe, D.J. (2000). Detection of intracellular oligomers of amyloid β-proteins in cells derived from human brain. *Biochemistry* 39, 10831-10839.

Walsh, D.M., Klyubin, I., Fadeeva, J.V., Cullen, W.K., Anwyl, R., Wolfe, M.S., Rowan, M.J., Selkoe, D.J. (2002). Naturally secreted oligomers of amyloid beta proteins potently inhibit hippocampal long term potentiation in vivo. *Nature* 416, 535-539.

Wang, H.Y., Lee, D.H., D'Andrea, M.R., Peterson, P.A., Shank R.P., Reitz, A.B. (2000). Beta-Amyloid(1-42) binds to alpha7 nicotinic acetylcholine receptor with high affinity. Implications for Alzheimer's disease pathology. *J. Biol. Chem.* 275, 5626-5632.

Wang, H.W., Pasternak, J.F., Kuo, H., Ristic, H., Lambert, M.P., Chromy, B., Viola, K.L., Klein, W.L., Stine, W.B., Krafft, G.A., Trommer, B.L. (2002). Soluble oligomers of β amyloid (1-42) inhibit long-term potentiation but not long-term depression in rat dentate gyrus. *Brain Res.* 924, 133-140.

Wang, J.Y., Shum, A.Y., Ho, Y.J. (2003). Oxidative neurotoxicity in rat cerebral cortex neurons: synergistic effects of H2O2 and NO on apoptosis involving activation of p38 mitogen-activated protein kinase and caspase-3. *J. Neurosci. Res.* 72, 508-519.

Wang, Q., Walsh, D.M., Rowan, M.J., Selkoe, D.J., Anwyl, R. (2004). Block of LTP induction in vitro by cell-derived and synthetic Aβ is mediated via activation of the kinases JNK, Cdk5 and p38 MAPK, and mGluR5. *J. Neurosci.* 24, 6047-6056.

Waters, D.L., Yau, C.L., Montoya, G.D., Baumgartner, R.N. (2003). Serum sex hormones, IGF-1, and IGFBP3 exert a sexually dimorphic effect on lean body mass in aging. J. Geronyol. *A Biol. Sci. Med. Sci.* 58, 648-652.

Weggen, S., Eriksen, J.L., Das, P., Sagi, S.A., Wang, R., Pietrzik, C.U., Findlay, K.A., Smith, T.E., Murphy, M.P., Bulter, T., Kang, D.E., Marquez-Sterling, N., Golde, T.E., Koo, E.H. (2001). A subset of NSAIDs lower amyloidogenic Abeta42 independently of cyclooxygenase activity. *Nature* 414, 212-216.

Wegiel, J., Wang, K.C, Tarnawski, M., Lach, B. (2000). Microglial cells are the driving force in fibrillar plaque formation, whereas astrocytes are a leading factor in plaque degradation. *Acta Neuropathol.* (Berl.) 100, 356-364.

Wegiel, J., Wang, K.C., Imaki, H., Rubenstein, R., Wronska, A., Osuchowski, M., Lipinski, W.J., Walker, L.C., LeVine, H. (2001). The role of microglial cells and astrocytes in fibrillar plaque evolution in transgenic APP(SW) mice. *Neurobiol. Aging.* 22, 49-61.

Wei, T., Chen, C., Hou, J., Xin, J., Mori, A. (2000). Nitric oxide induces oxidative stress and apoptosis in neuronal cells, Biochim. *Biophys. Acta.* 1498, 72-79.

Weldon, D.T., Rogers, S.D., Ghilardi, J.R., Finke, M.P., Clearly, J.P., O'Hare, E., Esler, W.P., Maggio, J.E., Mantyh, P.W. (1998). Fibrillar β-amyloid induces microglial phagocytosis, expression of inducible nitric oxide synthase, and loss of a select population of neurons in the rat CNS in vivo, *J. Neurosci.* 18, 2161-2173.

Wen, Y., Onyewuchi, O., Yang, S., Liu, R., Simpinks, J.W. (2004). Increased beta-secretase activity and expression in rats following transient cerebral ischemia. *Brain Res.* 1009, 1-8.

Wesselingh, S.L., Gough, N.M., Finlay-Jones, J.J., McDonald, P.J. (1990). Detection of cytokine mRNA in astrocyte cultures using the polymerase chain reaction. *Lymphokine Res.* 9, 177-185.

Westphalen, R.I., Scott, H.L., Dodd, P.R. (2003). Synaptic vesicle transport and synaptic membrane transporter sites in excitatory amino acid nerve terminals in Alzheimer disease. *J. Neural Transm.* 110, 1013-1027.

Willenborg, D.O., Fordham, S., Bernard, C.C., Cowden, W.B., Ramshaw, I.A. (1996). IFN-gamma plays a critical down-regulatory role in the induction and effector phase of mylin oligodendrocyte glycoprotein-induced autoimmune encephalomyelitis. *J. Immunol.* 157, 3223-3227.

Withson, J.S., Selkoe, D.J., Cotman, C.W. (1989). Amyloid beta protein enhances the survival of hippocampal neurons in vitro. *Science* 243, 1488-1490.

Wisniewski, T., Ghiso, J., Frangione, B. (1997). Biology of Aβ amyloid in Alzheimer's disease. *Neurobiol. Dis.* 4, 313-328.

Witting, A., Muller, P., Hermann, A., Kettenmann, H., Nolte, C. (2000). Phagocytic clearance of apoptotic neurons by Microglia/Brain macrophages in vitro: involvement of lectin-, integrin-, and phosphatidyl-serine-mediated recognition. *J. Neurochem.* 75, 1060-1070.

Wolf, B.B., Lopes, M.B., Vanden-Berg, S.R., Gonias, S.L. (1992). Characterization and immunohistochemical localization of alpha 2-macroglobulin receptor (low-density lipoprotein receptor-related protein) in human brain. *Am. J. Pathol.* 141, 37-42.

Wolozin, B., Kellman, W., Rousseau, P., Celesia, G.G., Sieguel, G. (2000). Decreased prevalence of Alzheimer disease associated with 3-hydroxy-3-methyglutaryl coenzyme A reductase inhibitors. *Arch. Neurol.* 57, 1439-1443.

Wood, W.G., Schroeder, F., Igbavboa, U., Avdulov, N.A., Chochina, S.V. (2002). Brain membrane cholesterol domains, aging and amyloid beta-peptides. *Neurobiol. Aging* 23, 685-694.

Wyss-Coray, T., Mucke, L. (2002) Inflammation in neurodegenerative disease -a double-edged sword. *Neuron* 35, 419-432.

Wyss-Coray, T., Yan, F., Lin A.H., Lambris, J.D., Alexander, J.J., Quigg, R.J., Masliah, E. (2002). Prominent neurodegeneration and increased plaque formation in complement-inhibited Alzheimer's mice. *Proc. Natl. Acad. Sci. USA* 99, 10837-10842.

Wyss-Coray, T., Loike, J.D., Brionne, T.C., Lu, E., Anankov, R., Yan, F., Silverstein, S.C., Husemann, J. (2003). Adult mouse astrocytes degrade amyloid-β in vitro and in situ. *Nature Med.* 9, 453-457.

Xiao, B., Bai, X., Zhang, G., Höjeberg, B., Link, H. (1996). Shift from anti- to pro-inflammatory cytokine profiles in microglia through LPS- or IFN-γ-mediated pathways. *NeuroReport* 7, 1893-1898.

Xie, L., Helmerhorst, E., Taddei, K., Plewright, B., Van Bronswijk, W., Martins, R. (2002). Alzheimer's beta-amyloid peptides compete for insulin binding to the insulin receptor. *J. Neurosci.* 22, RC221: 1-5.

Yaar, M., Zhai, S., Pilch, P.F., Doyle, S.M., Eisenhauer, P.B., Fine, R.E., Gilchrest, B.A. (1997). Binding of beta-amyloid to the p75 neurotrophin receptor induces apoptosis. A possible mechanism for Alzheimer's disease. *J. Clin. Invest.* 100, 2333-2340.

Yaar, M., Zhai, S., Fine, R.E., Eisenhauer, P.B., Arble, B.L., Stewart, K.B., Gilchrest, B.A. (2002). Amyloid beta binds trimers as web as monomers of the 75-kDa neurotrophin receptor and activates receptor signaling. *J. Biol. Chem.* 277, 7720-7725.

Yamamoto, T., Maruyama, W., Kato, Y., Yi, H., Shamoto-Nagai, M., Tanaka, M., Sato, Y., Naoil, M. (2002). Selective nitration of mitochondrial complex I by peroxynitrite: involvement in mitochondria dysfunction and cell death of dopaminergic SH-SY5Y cells. *J. Neural Transm.* 109, 1-13.

Yan, S.D., Chen, X., Fu, J., Chen, M., Zhu, H., Roher, A., Slattery, T., Zhao, L., Nagashima, M., Morser, J., Migheli, A., Nawroth, P., Stern, D., Schmidt, A.M. (1996). RAGE and amyloid-beta peptide neurotoxicity in Alzheimer's disease. *Nature* 382, 685-691

Yankner, B.A., Duffy, L.K., Kirschner, D.A. (1990). Neurotrophic and neurotoxic effects of amyloid beta protein: reversal by tachykinin neuropeptides. *Science* 250, 279-282.

Yao, P.J., Zhu, M., Pyun, E.I., Brooks, A.I., Therianos, S., Meyers, V.E., Coleman, P.D. (2003). Defects in expression of genes related to synaptic vesicle trafficking in frontal cortex of Alzheimer's disease. *Neurobiol. Dis.* 12, 97-109.

Yasojima, K., Schwab, C., McGeer, E.G., McGeer, P.L. (2000). Human neurons generate C-reactive protein and amyloid P: upregulation in Alzheimer's disease. *Brain Res.* 887, 80-89.

Yip, C.M., Elton, E.A., Darabie, A.A., Morrison, M.R., McLaurin, J. (2001). Cholesterol, a modulator of membrane-associated Abeta-fibrillogenesis and neurotoxicity. *J. Mol. Biol.* 311, 723-734.

Yokota, M., Saido, T.C., Tani, E., Yamaura, I., Minami, N. (1996). Cytotoxic fragment of amyloid precursor protein accumulates in hippocampus after global forebrain ischemia. *J. Cereb. Blood. Flow Metab.* 16, 1219-1223.

Youkin, S.G. (2001). Amyloid beta vaccination: reduced plaques and improved cognition. *Nat. Med.* 7, 18-19.

Zhang, Z., Chopp, M., Powers, C. (1997a). Temporal profile of microglial response following transient (2h) middle cerebral artery occlusion. *Brain Res.* 744, 189-198.

Zhang, F., Eckman, C., Youkin, S., Hsiao, K.K. Iadecola, C. (1997b). Increased susceptibility to ischemic brain damage in transgenic mice overexpressing the amyloid precursor protein. *J. Neurosci.* 17, 7655-7661.

Zhang, Y., Hong, Y., Bounhar, Y., Blacker, M., Roucou, X., Tounekti, O., Vereker, E., Bowers, W.J., Federoff, H.J., Goodyer, C.G., LeBlanc, A. (2003). P75 neurotrophin receptor protects primary cultures of human neurons against extracellular amyloid beta peptide cytotoxicity. *J. Neurosci.* 23, 7385-7394.

Zhu, Y., Roth-Eichhorn, S., Braun, N., Culmsee, C., Rami., A., Krieglstein, J. (2000). The expression of transforming growth factor-beta 1 (TGF-β1) in hippocampal neurons: a temporary upregulated protein level after transient forebrain ischemia in the rat. *Brain Res.* 866, 286-298.

Zhu, Y., Yang, G.Y., Ahlemeyer, B., Pang, L. Che, X.M., Culmsee, C. (2002). Transforming growth factor-β1 increases bad phosphorylation and protects neurons against damage. *J. Neurosci.* 22, 3898-3909.

Zhu, Y., Culmsee, C., Klumpp, S., Krieglstein, J. (2004a). Neuroprotection by transforming growth factor-β1 involves activation of nuclear factor-κB through phosphatidylinositol-

3-OH kinase/Akt and mitogen-activated protein kinase-extracellular-signal regulated kinase1, 2 signaling pathways. *Neurosci.* 123, 897-906.

Zhu, X., Raina, A.K., Perry, G., Smith, M.A. (2004b). Alzheimer's disease: the two-hit hypothesis. *Lancet Neurol.* 3, 219-226.

Zou, K., Cong, J.S., Yaganisawa, K., Michikawa, M. (2002). A novel function of monomeric amyloid beta-protein serving as an antioxidant molecule against metal-induced oxidative damage. *J. Neurosci.* 22, 4833-4841.

In: Focus on Brain Research
Editor: Casey J. Resch, pp. 87-105
ISBN 1-600021-395-2

Chapter III

The Blood Brain Barrier: An Introduction with an Historical Overview

Eve M. Taylor[*]
ACADIA Pharmaceuticals Inc. San Diego, USA.

Introduction

The blood-brain barrier (BBB) creates and maintains the highly regulated extracellular environment of the CNS through three "lines of defense": - (1) a physical barrier formed by tight junctions between endothelial cells of the brain capillaries and epithelial cells of both the choroid plexus and the arachnoid membrane (Figure 1), (2) transporters that mediate the efflux of compounds from brain to blood, and (3) an enzymatic barrier conferred by an enrichment of degradative enzymes in these endothelial and epithelial cells. A perusal of 20th and 21st century literature on the BBB quickly reveals that the major focus of BBB research has been the first of these components; this is particularly highlighted by the fact that much of the effort to develop drugs that can effectively distribute into the CNS has been based on conquering this component of the barrier. While several drug delivery strategies have been developed including the osmotic opening of the BBB and vector-based transport mechanisms (Madrid et al., 1991; Pardridge, 1998, 2002), optimization of the physicochemical characteristics of drugs has been the primary focus of the pharmaceutical industry when "designing" drugs which will penetrate the BBB; indeed the octanol:water partition coefficient (log P) is often used as a primary screen in the drug discovery process (Li and Kerns, 2003). However, upon examination of the correlation between lipophilicity and BBB penetration it is apparent that there are several compounds for which the correlation between these two measures is poor and they do not penetrate the brain to the level predicted by logP

(Banks, 1985; Pardridge, 1998; Golden and Pollack, 2003). In many cases this is likely because these compounds are transported from brain to blood by efflux transporters. The goal of this book is to provide a comprehensive review of the information available on efflux transport across the blood-brain barrier and to examine the role of this process in the preservation of CNS homeostasis with respect to both endogenous and exogenous substances.

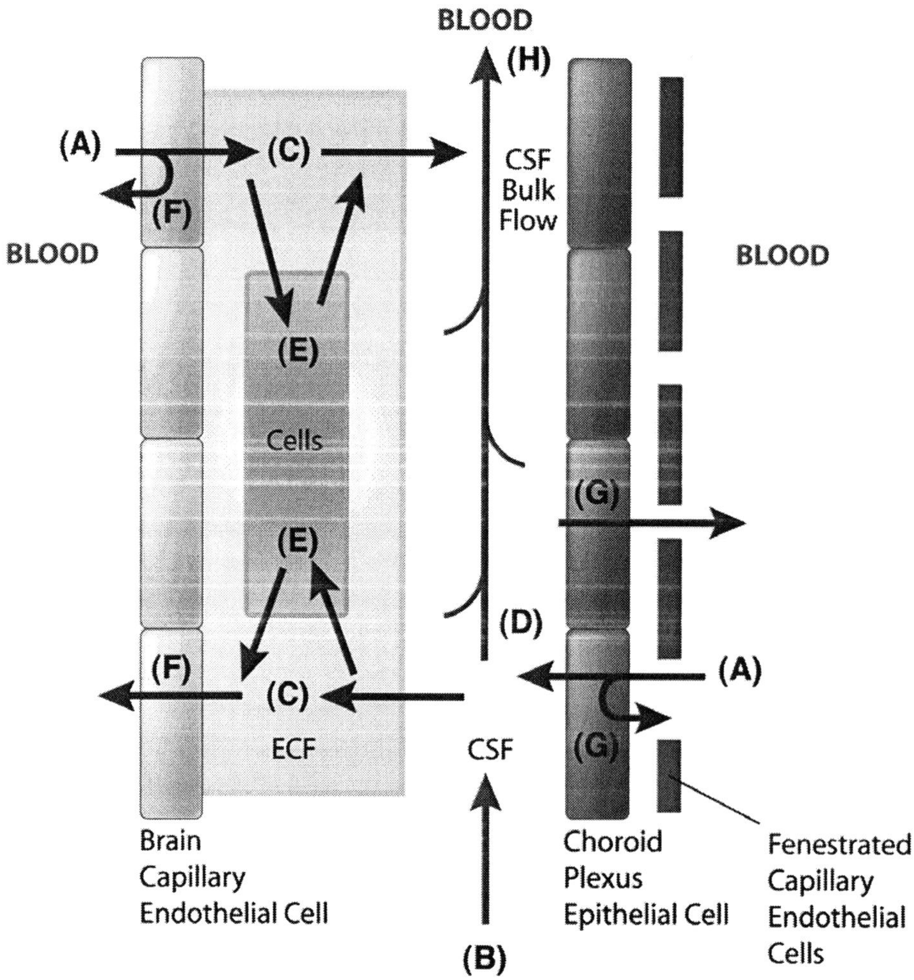

Figure 1. Sites of the Blood-brain Barrier. The blood-brain barrier comprises tight junctions between (1) the endothelial cells of brain capillaries, (2) epithelial cells of the choroid plexus, and (3) epithelial cells of the arachnoid membrane. Reproduced from Taylor (2002) with permission from Adis International.

* Correspondence concerning this article should be addressed to Eve M. Taylor, ACADIA Pharmaceuticals Inc. 3911 Sorrento Valley Road, San Diego, CA92121. USA.

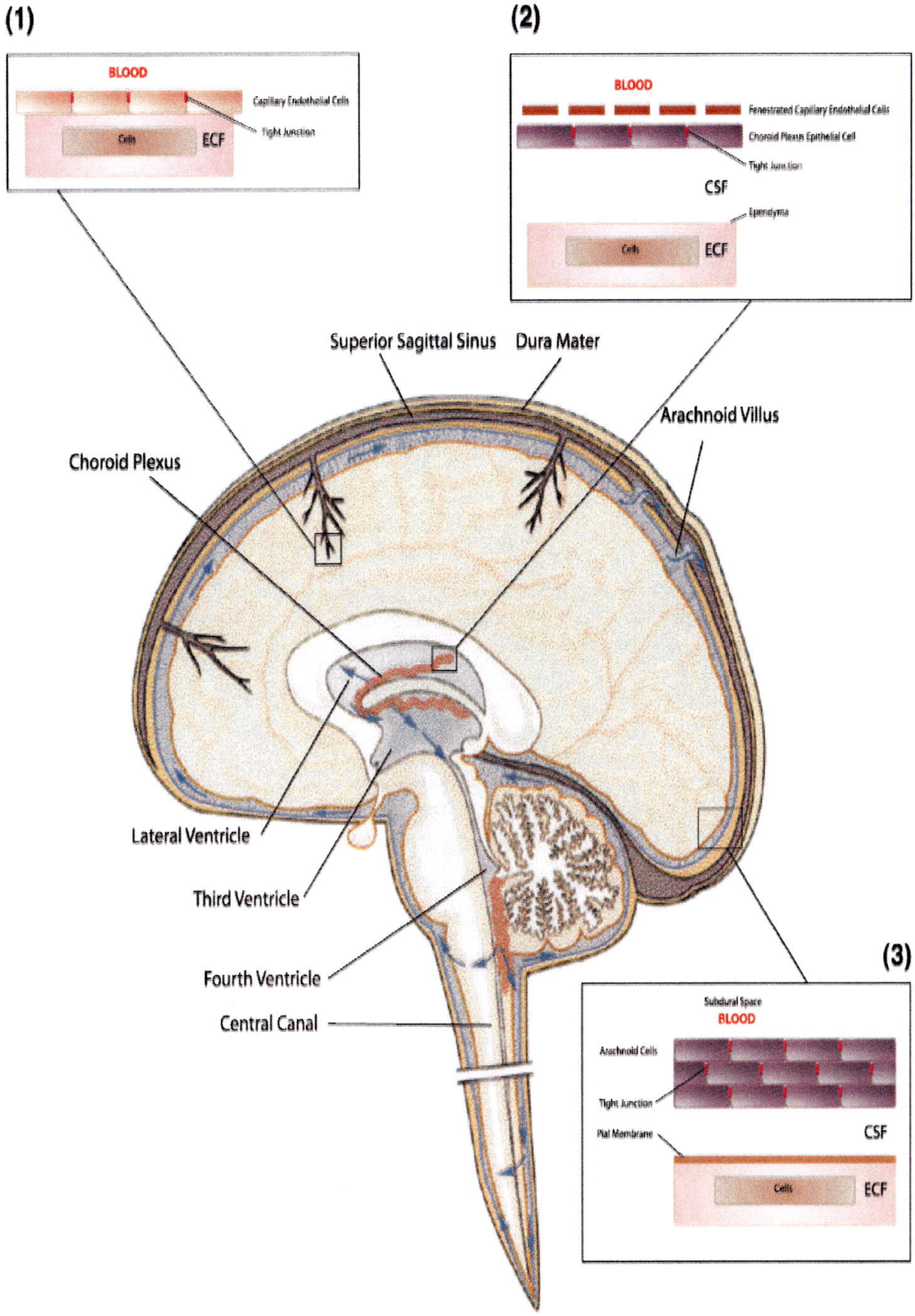

Figure 2. Pathways of Drug Distribution in Brain. Once a drug enters the brain by systemic administration (A), or direct delivery into the CNS (B), it will become distributed in the extracellular fluid (ECF) of the brain parenchyma (C) and the cerebrospinal fluid (CSF; D) and may interact with the cells of the brain (E). It may then return to the blood by crossing the brain capillary endothelial cells (F) or the choroids plexus epithelial cells (G), or by bulk flow of CSF through arachnoid villi (H). Reproduced from Taylor (2002) with permission from Adis International.

Passages of Efflux from Brain

In 1914 Dandy and Blackfan demonstrated that phenolsulfonphthalein (PSP) appeared in the urine within minutes of injection into the cisterna magna and, in spite of the rapidity with which this compound left the CNS, they concluded that PSP was carried into the peripheral circulation by bulk flow of cerebrospinal fluid (CSF) and passage through the arachnoid villi. This assumption went unchallenged for nearly 50 years and was the basis of Dandy's PSP test for ventriculocisternal block, a clinical tool used to diagnose hydrocephalus (REF). In 1961 Pappenheimer and colleagues suggested that PSP clearance from the CNS may in fact occur through active processes. These workers showed that phenosulfonphthalein and iodopyracet were transported from CSF to blood and that the transport of iodopyracet was inhibited by phenosulfonphthalein and *p*-aminohippurate. Subsequently, the saturable disappearance from brain of other organic anions (Prockop et al., 1962), organic cations (Schanker et al., 1962), iodide (Coben et al., 1964; Davson and Pollay, 1964; Reed et al., 1965), potassium (Cserr, 1965; Katzman et al., 1965), and penicillin (Fishman, 1966; Dixon et al., 1969) was demonstrated. It is now known that electrolytes, peptides, hormones, neurotransmitters, neuromodulators, and cytokines are transported out of the CNS by active efflux (Banks, 1999; Chapters 2 and 3) and that such processes are an integral component of BBB function.

The majority of early in vivo studies employed the technique of ventriculocisternal perfusion in which drug is perfused into the cerebroventricular system via the cisterna magna thus exposing the choroid plexus epithelium and the ependymal lining of the ventricles to the compound of interest. Investigators concluded from these early experiments that the primary site of active efflux mechanisms was the choroid plexus and this conclusion was supported by in vitro experiments (Cameron, 1953; Becker 1961; Rall and Sheldon, 1962; Welch, 1962a, b). The CP indeed plays a major role in the maintenance of CNS homeostasis via efflux and in Chapter 4 this role is discussed in detail. Additional techniques to evaluate efflux from brain have been developed over the years (Chapters 2 and 3) and have facilitated the demonstration of efflux mechanisms at the brain capillary endothelium. Thus it is now clear that once a compound enters the brain, after systemic administration or direct administration into the CNS, it may return to the blood by the passive mechanisms of CSF bulk flow (with reabsorption at the arachnoid villi) and diffusion across the BBB and BCSFB or by active efflux mechanisms located at both of the brain-blood interfaces (Figure 2).

Molecular Identity of Active Efflux at the BBB

During the 1960s, 1970s and much of the 1980s the field of efflux transport across the blood-brain barrier was largely observational with many studies published demonstrating that saturable efflux did occur for a number of different compounds. The turning point came in the late 1980s when Cordon-Cardo et al. (1989) and Thiebaut et al. (1989) showed that the efflux transporter P-glycoprotein (P-gp) is expressed in capillary endothelial cells in both rats and humans. P-gp is the prototypical multidrug resistance (MDR) transporter which was originally identified in drug resistant cancer cells (Gottesman and Pastan, 1993; Gottesman et al., 1996). The expression of this transporter at the BBB has now been confirmed in tissue sections, isolated brain capillaries, isolated luminal membranes from brain capillaries, primary cultures of brain capillary endothelial cells, and in immortalized cultures of brain capillary endothelial cells (Tishler et al., 1995; Sugawara et al., 1990; Stewart et al., 1996; Schumacher et al., 1997; Jette et al., 1993, 1995; Seetharaman et al., 1998; Matsuoka et al., 1999; Regina et al., 1998; Qin et al., 1995; Beaulieu et al., 1995, 1997; Barrand et al., 1995; Begl;ey et al., 1996; Bradley et al., 1990; Cordon-Cardo et al., 1990; Cropp et al., 1989; Hegmann et al., 1992; Decleves et al., 2000; Lechardeur et al., 1995; Tatsuta et al., 1992; Mercier et al., 2004). Controversy has arisen around the findings of Pardridge and coworkers (Pardridge et al., 1997; Golden and Pardridge, 2000) who have asserted that P-gp is expressed in astrocytic end feet and not endothelial cells. The most recent investigation of P-gp localization within the brain suggests that the transporter may be expressed in both locations (Mercier et al., 2004). Furthermore, P-gp has also been found in choroid plexus in rats (Rao et al., 1999) and monkeys (Warren et al., 2000).

In 1992 the first studies demonstrating the functionality of P-gp in brain capillary endothelial cells were published (Tatsuta et al., 1992; Tsuji et al., 1992), however it was gene knockout studies that provided the most convincing experimental description of a role for P-gp in the BBB. Brain penetration of a wide range of drugs is greatly increased in mdr1a knockout mice (Schinkel et al., 1994, 1995, 1996; Kim et al., 1998; Mayer et al., 1996; Murata et al., 1999) and the functional implication of this observations is clearly illustrated by the fact that ivermectin, commonly used to treat infections and known for its high safety margin in vertebrates, is lethally neurotoxic in mdr1a knockout animals (Schinkel, 1999). The biology of P-gp in the BBB is discussed in Chapter 3.

Multidrug resistance-associated protein (MRP) 1 was the first non-P-gp MDR-conferring transporter identified (Borst et al., 1999). It transports a wide range of compounds, many of which are actively transported out of brain. The initial examination of MRP1 expression in human tissues demonstrated that MRP1 is expressed in brain (Zaman et al., 1993) and, in several other species, studies using isolated microvessels, tissue sections, and cultured brain capillary endothelial cells have indicated that MRP1 is expressed at the BBB (Seetharaman et al., 1998; Gutmann et al., 1999; Huai-Yun et al., 1998; Kusuhara et al., 1998; Decleves et al., 2000; Regina et al., 1998; Zhang et al., 2000; Stride et al., 1996; Miller et al., 2000). Additionally, Rao et al. (1999) and Nishino et al. (1999) have demonstrated MRP1 in the choroid plexus from rat and Mercier et al. (2004) have shown expression in astrocytes. MRP1-deficient mice have been generated and no BBB-related deficits have been detected in these animals (Wijnholds et al., 1997, 1998; Lorico et al., 1997). However, measurement of etoposide penetration into brain in mdr1a/mdr1b double knock outs (DKOs) and mrp1/mdr1a/mdr1b triple knockouts (TKOs) showed that TKOs had increased CSF etoposide levels compared to DKOs indicating that MRP1 may play a role specifically at the BCSFB (Wijnholds et al., 2000). Recently, MRP1 homologs MRP2-6 have been identified (Kool et al., 1997, 1999; McAleer et al., 1999) and all have been shown to mediate drug efflux in vitro. Analysis of human brain tissue by Northern blot and RNAse protection assay demonstrated that MRP5 was highly expressed in brain (Kool et al., 1997; McAleer et al., 1999). More recently, Miller et al. (2000) detected MRP2 in capillary endothelial cells of isolated rat and porcine brain capillaries and Zhang et al. (2000) demonstrated expression of

MRP4, 5 and 6 in cultured bovine brain capillary endothelial cells and isolated bovine brain capillaries. The potential role of MRPs in the efflux of substrates from CNS to blood is discussed in Chapter 6.

Yet another MDR transporter may play a role in the efflux of compounds from the brain into the blood. Breast cancer resistance protein (BCRP) was identified in chemotherapeutic drug-resistant cell lines however, like P-gp, it is also expressed in non-malignant tissues including brain capillary endothelial cells and may thus act to restrict entry of potentially toxic substrates into brain. The information available on BCRP and how it may function in the brain is discussed in detail in Chapter 7.

Monocarboxylic acid transporters (MCTs) have been implicated in both the influx (Drewes et al., 1999) and efflux of compounds across the blood-brain barrier. Probenecid (Deguchi et al., 1997), 6-mercaptopurine (Deguchi et al., 2000), aluminum citrate (Ackley and Yokel, 1997, 1998) and AIT-082 (Yan et al., 2002), are transported out of brain by a mechanism that can be inhibited by monocarboxylic acids and MCTs have been implicated in each case. MCT1 is expressed in brain capillary endothelial cells, ependymocytes, astrocytic endfeet, pericytes and choroid plexus (Koehler-Stec et al., 1998; Gerhart et al., 1997; Takanaga et al., 1995; Leino et al., 1999) and MCT2 is abundant in astrocytic endfeet, ependymocytes and neuropil (Gerhart et al., 1998; Jackson et al., 1997). An additional six MCTs (MCT3-8) have been cloned from human tissue and both MCT7 and 8 are expressed in brain (Halestrap and Price, 1999; Lafreniere et al., 1994; Price et al., 1998) however their role in BBB is yet to be determined. MCT transporters are discussed in Chapter 8.

Four major families of organic ion transporters have been identified; Organic Anion Transporters (OAT), Organic Cation Transporters (OCT), organic anion transporter proteins (oatp), and Organic Cation/Carnitine Transporters (OCTN) (Koepsell, 1998; Inui et al, 2000; Sekine et al., 2000). Multiple members of each family have been identified (OAT1-4, OCT 1-3, oatp 1-4, and OCT/OCTN 1-3) and many of these transporters have substrate profiles that include compounds that are actively transported out of brain (Taylor, 2002). There is evidence that OAT1 and 3 (Kusuhara et al., 1999; Hosoyamada et al., 1999), OCT1-3 (Zhang et al., 1997; Busch et al., 1998; Grundemann et al., 1997; Kekuda et al., 1998; Wu et al., 1998), oatp1-3 (Angeletti et al., 1997, 1998; Noe et al., 1997; Abe et al., 1998; Asaba et al., 2000; Walters et al., 2000), and OCTN1-3 (Tamai et al., 1997, 1998, 2000; Wu et al., 1999, 2000) are expressed in brain and their role in brain-to-blood efflux is under investigation. To date, the saturable efflux of DHEAS, an oatp2 substrate, has been demonstrated in immortalized mouse brain capillary endothelial cells (Asaba et al., 2000) and 17β-estradiol 17β-D-glucuronide (E217βG), an oatp substrate, was transported out of brain by a probenecid-sensitive mechanism (Nishino et al., 1999). These transporters are discussed in Chapters 3, 4, 6, and 9.

In addition to these transporters which have been cloned, there are a number of other efflux transporters which have been pharmacologically identified. Many of these are outlined in the chapters 2 and 3 and mediate the efflux of endogenous substances including electrolytes, hormones, peptides, neurotransmitters, neuromodulators and cytokines. Thus BBB efflux not only serves to expel exogenous compounds that have made it across the physical BBB, but also maintain CNS homeostasis by excreting by-products of central metabolism. Molecular characterization of these transporters will greatly enhance the progress of this field and, given the increased focus of the pharmaceutical industry on biologic drugs including peptides, cloning of peptide efflux transporters in particular may provide valuable insight into more effective delivery of such drugs to the CNS.

Clinical Impact of Efflux Transporters in the BBB

The discovery in 1961 by Pappenheimer and coworkers that active efflux of compounds out of the brain can occur had an immediate impact in the clinic. The original observation made by Dandy and Blackfan in 1914 that PSP rapidly moved out of the CNS had been translated into a diagnostic test for hydrocephalus. With the subsequent findings of Pappenheimer et al. (1961) it became apparent that the active efflux of PSP could influence this test and generate artifactual results and thus potentially incorrect diagnoses. Given the large number of diagnostic tests that require CNS penetration of tracers (e.g., PET) the role that efflux may play in the distribution of any tracer should be considered (Waterhouse, 2003).

In 1948 Boger et al. reviewed several years of observations that revealed the poor penetration of drugs including penicillin into the brain and it is now a certainty that the minimal effectiveness, in some patients, of antibiotics on CNS infections, anticonvulsants on epilepsy, opiates on CNS-based pain, chemotherapy on brain tumors, and more recently HIV drugs on AIDS dementia, may be explained, at least in part, by the efficiency with which efflux transporters remove these compounds from the CNS (Adkison et al., 1994, 1996; Cornford et al., 1985; Deguchi et al., 1997, 2000; Doze et al., 2000; Galinsky et al., 1991; Hedaya et al., 1989; Ooie et al, 1997; Spector and Lorenzo, 1974; Susuki et al., 1989; Wang et al., 1995).

Changes in the levels of and/or activity of efflux transporters occur with age, in neurological disorders, and because of P-gp substrates themselves (Aggarwal et al., 1997; Thiebaut et al., 1989; Lee and Bendayan, 2004). There is, for example, evidence that there is an increased expression of P-gp, MRP1, and MRP2 in patients with refractory epilepsy (Tishler et al., 1995; Aronica et al., 2004) which likely increases the efflux of antiepileptic medication thus reducing brain concentrations. On the other hand, impairment in peptide efflux mechanisms may decrease efflux of methionine enkephalin which is known to play a role in alcoholism (Chapter 2) and in the accumulation of amyloid peptide in Alzheimer's disease which may contribute to the deposition of amyloid plaques (see Chapter 2). It remains to be shown whether such changes in the expression of P-gp are etiological or pathological. With regard to etiology, genetic research has recently revealed the association of P-gp polymorphisms with Alzheimer's disease, epilepsy, and Parkinson's disease (Marzolini et al., 2004; Lee and Bendayan, 2004, Siddiqui et al., 2003; Zimprich et al., 2004).

Given the clinical impact that BBB efflux mechanisms have on CNS function and the treatment of CNS disorders, modulation of such mechanisms may be of value. The concept of modulating efflux mechanisms to increase brain penetration of drugs was introduced in the 1961 study by Pappenheimer and coworkers. They showed that the efflux of iodopyracet from CSF was inhibited by phenolsulfonphthalein and by *p*-aminohippurate and suggested that inhibition of efflux mechanisms might enhance brain penetration of drugs such as penicillin that had been shown to penetrate brain poorly. Indeed, Fishman (1966), Dixon et al. (1969) and Spector and Lorenzo (1974) went on to show that probenecid, a drug commonly used to inhibit renal clearance of antibiotics, increased the CSF levels of penicillin. In more recent studies the efflux of dideoxyinosine, AZT, γ-aminobutyric acid, 6-mercaptopurine, leukotriene C_4, valproic acid, and etoposide has been inhibited by co-administration of a wide range of substances used as putative specific inhibitors of efflux transporters (Burgio et al., 1996; Adkison et al., 1994; Doze et al., 2000; Takasawa et al., 1997; Hedaya et al., 1989; Wong et al., 1993; Wang et al., 1997; Dykstra et al., 1993; Spector and Goetzel, 1986; Loscher and Frey, 1982; Deguchi et al., 1997, 2000). It is apparent however, that there is extensive overlap of the substrate profiles of all the transporters discussed here (Taylor, 2002) and therefore specific transport inhibitors are required. Several companies have attempted to develop specific efflux inhibitors (Taylor, 2002), the most well characterized of which is valspodar (PSC 833), a non-immunosuppressive cyclosporin A analog that specifically inhibits P-gp. Its utility in the inhibition of drug efflux from brain has been investigated and it has been shown to increase the uptake of colchicine, vinblastine, digoxin, quinidine, vincristine and cyclosporin A into brain (Drion et al., 1996; Desrayaud et al., 1997; Lemaire et al., 1996; Mayer et al., 1997; Kusuhara et al., 1997) and increase neuro-sensitivity to cyclosporin A, quinidine, and ivermectin (Didier et al., 1995).

In addition to modulating efflux transporters with small molecule pharmacological agents, it may be possible to use anti-transporter antibodies, protein kinase C inhibitors, antisense oligonucleotides, or ribozymes to down-regulate the activity or expression of transporters (Volm et al., 1998; Sonneveld et al., 1997). However, the very nature of many efflux mechanisms (i.e., high capacity and low specificity) means that inhibition of such mechanisms to enhance BBB penetration will not be specific to a single drug and will enhance transport of many compounds into the brain. This may have dire consequences for patients, particularly those on multiple therapies. Design of drugs that have reduced affinity for brain efflux transporters or that can bypass the transporters may provide a more efficient solution to the problem of efflux in the BBB. Susuki and coworkers (1989) have demonstrated that cephalosporin antibiotics including ceftriaxone and imipenem have minimal affinity for the efflux system that is responsible for transporting penicillins out of brain and increased brain penetration of these drugs is associated with improved treatment of bacterial meningitis. Systematic investigations of the structure-activity relationship of efflux transporter substrates may improve the effectiveness of this approach. However, strategies that shield compounds from efflux may be the most likely to succeed. Such strategies are drug-specific and thus will not block the efflux of other compounds which could cause neurotoxicity and designing molecules which retain pharmacological activity and yet lose efflux transporter substrate status is difficult. "Shielding" strategies include conjugation of drugs to peptides that readily cross the BBB (Rousselle et al., 2000), encapsulation of drugs within nanoparticles (Lockman et al., 2003; Koziara et al., 2003, 2004), and immunoliposomes that are coated with proteins (e.g., tranferrin) that are transported into brain by specific receptors (Pardridge, 2002). These strategies are discussed in Chapter 3.

Conclusion and Looking to the Future

With the unequivocal demonstration, in P-gp knockout mice, of a role for drug efflux in the preservation of the CNS environment, an area of research that dates back to the early 1960's is gaining momentum. The complexity of the field is well illustrated by the recent paper by Choudhuri and colleagues (2003) in which they demonstrated that mRNA for Mdr1a, Mdr1b, Mdr2, Mrp1, Mrp2, Mrp3, Mrp4, Mrp5, Oatp1, Oatp2, Oatp3, Oatp4, Oatp5, Oatp12, Oct1, Oct2, Menke's transporter, DMT1, Ent1, Ent 2, Pept1, Pept2, OctN1 and N2, , Oat1, Oat2, Oat3, Cnt1, Cnt 2, Oat-K (1/2), Ntcp, Bsep, Ibat, AbcG5, AbcG8 transporters were all detected in rat choroid plexus. It is yet to be determined whether all of these transporters are expressed at the protein level and what their role is in regulation of the CNS environment. It is however, clear that the number of transporters expressed in the BBB and BCSFB is extensive. Many of these transporters are expressed in the periphery and operate in hepatic and renal clearance of toxic and potentially toxic compounds. It has been known for more than 100 years that renal excretion of organic cations is an active process (Weiner and Mudge, 1964) and the early investigators of brain-to-blood efflux were quick to draw parallels between efflux mechanisms observed in the BBB and those seen in the kidney. Indeed the identification of transporters in the kidney and the liver has provided a valuable source of information which has been translated to the BBB. Furthermore, the mechanisms of action of the transporters, how they are regulated, how they interact with other drug clearing processes (e.g., enzymatic degradation of drugs) is largely study in the context of these peripheral transporters. The lessons learnt in the periphery will no doubt continue to impact our understanding of efflux transporters in the BBB and this is discussed in chapter 9.

The goal of this book is to provide readers with a comprehensive review of the state-of-the-art of efflux transport across the blood-brain barrier and to provide detailed "food-for-thought" which will hopefully stimulate more research in this area in both the academic and industrial research communities. It is clear from the ever burgeoning collection of publications in this area that the study of efflux transporters and the ability to distinguish between transporters with high capacity and low specificity will be a great challenge. However, with the increased genomic and proteomic capabilities of our age, the limit to deciphering this puzzle is time and effort. Hopefully this volume will inspire researchers to expend both.

Acknowledgments

The author would like to thank Dr. Timothy Mayall and Alexander Mayall for their patience and support during the writing of this chapter and the editing of this book.

REFERENCES

Abe T, Kakyo M, Sakagami H, Tokui T, Nishio T, Tanemoto M, Nomura H, Hebert SC, Matsuno S, Kondo H and Yawo H. (1998). Molecular characterization and tissue distribution of a new organic anion transporter subtype (oatp3) that transports thyroid hormones and taurocholate and comparison with oatp2. *J Biol Chem, 273*: 22395-22401.

Ackley DC and Yokel RA (1997). Aluminum citrate is transported from brain into blood via the monocarboxylic acid transporter located at the blood-brain barrier. *Toxicol, 120*: 89-97.

Ackley DC and Yokel RA (1998). Aluminum transport out of brain extracellular fluid is proton dependent and inhibited by mersalyl acid, suggesting mediation by the monocarboxylate transporter (MCT1). *Toxicol, 127*: 59-67.

Adkison KDK, Artru AA, Poweres KM and Shen DD (1994). Contribution of probenecid-sensitive anion transport processes at the brain capillary endothelium and choroid plexus to the efficient efflux of valproic acid from the central nervous system. *J Pharm Exp Therap, 268*: 797-805.

Adkison KDK, Powers KM, Artru AA and Shen DD (1996). Effect of para-aminohippurate on the efflux of valproic acid from the central nervous system of the rabbit. *Epilepsy Res, 23*: 95-104.

Aggarwal S, Tsuruo T and Gupta S (1997). Altered expression and function of P-glycoprotein (170 kDa) encoded by the MDR 1 gene, in T cell subsets from aging humans. *J Clin Immunol, 17*: 448-454.

Angeletti RH, Bergwerk AJ, Novikoff PM and Wolkoff AW (1998). Dichotomous development of the organic anion transport protein in liver and choroid plexus. *Am J Physiol, 275*: C882-C887.

Angeletti RH, Novikoff PM, Juvvadi SR, Fritschy JM, Meier PJ and Wolkoff AW (1997). The choroid plexus epithelium is the site of the organic anion transport protein in the brain. *Proc Natl Acad Sci USA, 94*: 283-286.

Aronica E, Gorter JA, Ramkema M, Redeker S, Ozbas-Gercerer F, van Vliet EA, Scheffer GL, Scheper RJ, van der Valk P, Baayen JC and Troost D (2004). Expression and cellular distribution of multidrug resistance-related proteins in the hippocampus of patients with mesial temporal lobe epilepsy. *Epilepsia, 45*: 441-451.

Asaba H, Hosoya K, Takanaga H, Ohtsuki S, Tamura E, Takizawa T and Terasaki T (2000). Blood-brain barrier is involved in the efflux transport of a neuroactive steroid, dehydroepiandrosterone sulfate, via organic anion transporting polypeptide 2. *J Neurochem, 75*: 1907-1916.

Banks WA (1985). Peptides and the blood-brain barrier: lipophilicity as a predictor of permeability. Bra*in Res Bull 15*: 287-292.

Banks WA (1999). Physiology and pathology of the blood-brain barrier: implication for microbial pathogenesis, drug delivery and neurodegenerative disorders. *J Neurovirol, 5*: 538-555.

Barrand MA, Robertson KJ and von Weikersthal SF (1995). Comparisons of P-glycoprotein expression in isolated rat brain microvessels and in primary cultures of endothelial cells derived from microvasculature of rat brain, epidymal fat pad and from aorta. *FEBS Letts, 374*: 179-183.

Beaulieu E, Demeule M, Pouliot JF, Averill-Bates DA, Murphy GF and Beliveau R (1995). P-gp of blood-brain barrier: cross-reactivity of Mab C219 with a 190kDa protein in bovine and rat isolated brain capillaries. *Biochim Biophys Acta, 1233*: 27-32.

Beaulieu E, Demeule M, Ghitescu L and Beliveau R (1997). P-glycoprotein is strongly expressed in the luminal membranes of the endothelium of blood vessels in the brain. *Biochem J, 326*: 539-544.

Becker B (1961). Cerebrospinal fluid iodide. *Am J Physiol, 201*: 1149-1151.

Begley DJ, Lechardeur D, Chen ZD, Rollinson C, Bardoul M, Roux F, Scherman D and Abbott NJ (1996). Functional expression of P-glycoprotein in an immortalized cell line of rat brain endothelial cells, RBE4. *J Neurochem, 67*: 988-995.

Boger WP, Baker RB and Wilson WW (1948). *Proc Soc Exp Biol Med, 68*: 101-

Borst P, Evers R, Kool M and Wijnholds J (1999). The multidrug resistance protein family. *Biochim Biophys Acta, 1461*: 347-357.

Bradley G, Georges E and Ling V (1990). Sex-dependent and independent expression of the p-glycoprotein isoforms in Chinese hamster. *J Cell Physiol, 145*: 398-408.

Burgio DE, Gosland MP and McNamara PJ (1996). Modulation effects of cyclosporine on etoposide pharmacokinetics and CNS distribution in the rat utilizing microdialysis. *Biochem Pharmacol, 51*: 987-992.

Busch AE, Karbach U, Miska D, Gorboulev V, Akhoundova A, Volk C, Arndt P, Ulzheimer JC, Sonders MS, Baumann C, Waldegger S, Lang F and Koepsell H (1998). Human neurons express the polyspecific cation transporter hOCT2, which translocates monoamine neurotransmitters, amantadine, and memantine. *Mol Pharm, 54*: 342-352.

Cameron G (1953). Secretory activity of the choroid plexus in tissue culture. *Anat Rec, 117*: 115-125.

Choudhuri S, Cherrington NJ, Li N and Klaassen CD (2003). Constitutive expression of various xenobiotic and endobiotic transporter mRNAs in the choroid plexus of rats. *Drug Met Disp, 31*: 1337-1345.

Coben LA, Loeffier JD and Elsasser JC (1964). Spinal fluid iodide transport in dog. *Am J Physiol, 206*: 1373-1378.

Cordon-Cardo C, O'Brien JP, Boccia J, Casals D, Bertino JR and Melamed MR (1990). Expression of the multidrug resistance gene product (P-glycoprotein) in human normal and tumor tissues. *J Histochem Cytochem, 38*: 1277-1287.

Cordon-Cardo C, O'Brien JP, Casals D, Rittman-Grauer L, Biedler JL, Melamed MR and Bertino JR (1989). Multidrug-resistance gene (P-glycoprotein) is expressed by endothelial cells at blood-brain barrier sites. *Proc Natl Acad Sci USA, 86*: 695-698.

Cornford EM, Diep CP and Pardridge WM (1985). Blood-brain barrier transport of valproic acid. *J Neurochem, 44:* 1541-1550.

Croop JM, Raymond M, Haber D, Devault A, Arceci RJ, Gros P and Housman DE (1989). The three mouse multidrug resistance (mdr) genes are expressed in a tissue-specific manner in normal mouse tissues. *Mol Cell Biol, 9*: 1346-1350.

Cserr H (1965). Potassium exchange between cerebrospinal fluid, plasma and brain. *Am J Physiol, 209:* 1219-1226.

Dandy WE and Blackfan KD (1914). *Am J Dis Child*, *8*: 406

Davson H and Pollay M (1964). Influence of various drugs on the transport of ^{131}I and PAH across the cerebrospinal fluid-blood barrier. *J Physiol*, *167*: 239-246.

Decleves X, Regina A, Laplanche JL, Roux F, Boval B, Launay JM and Scherrmann JM (2000). Functional expression of P-glycoprotein and multidrug resistance-associated protein (Mrp1) in primary cultures of rat astrocytes. *J Neurosci Res*, *60*: 594-601.

Deguchi Y, Nozawa K, Yamada S, Yokoyama Y and Kimura R (1997). Quantitative evaluation of brain distribution and blood-brain barrier efflux transport of probenecid in rats by microdialysis: possible involvement of the monocarboxylic acid transport system. *J Pharm Exp Therap*, *280*: 551-560.

Deguchi Y, Yokoyama Y, Sakamoto T, Hayashi H, Naito T, Yamada S and Kimura R (2000). Brain distribution of 6-mercaptopurine is regulated by the efflux transport system in the blood-brain barrier. *Life Sci*, *66*: 649-662.

Desrayaud S, Guntz P, Scherrmann JM and Lemaire M (1997). Effect of the p-glycoprotein inhibitor, SDZ PSC 833, on the blood- and brain pharmacokinetics of colchicine. *Life Sci*, *61*: 153-163.

Didier AD and Loor F (1995). Decreased biotolerability for ivermectin and cyclosporin A in mice exposed to potent P-glycoprotein inhibitors. *Int J Cancer*, *63*: 263-267.

Dixon RL, Owens ES and Rall DP (1969). Evidence of active transport of benzyl-14C-Penicillin from cerebrospinal fluid to blood. *J Pharm Sci*, *58*:1106-1109.

Doze P, Van Waarde A, Elsinga PH, Hendrikse NH and Vaalburg W (2000). Enhanced cerebral uptake of receptor ligands by modulation of P-glycoprotein function in the blood-brain barrier. *Synapse*, *36*: 66-74.

Drewes LR (1999). What is the blood-brain barrier? A molecular perspective. *Adv Exp Med Biol*, *474*: 111-122.

Drion N, Lemaire M, Lefauconnier JM and Scherrmann JM (1996). Role of p-glycoprotein in the blood-brain transport of colchicine and vinblastine. *J Neurochem*, *67*: 1688-1693.

Dykstra KH, Arya A, Arriola DM, Bungay PM, Morrison PF and Dedrick RL (1993). Microdialysis study of zidovudine (AZT) transport in rat brain. *J Pharm Exp Therap*, *267*: 1227-1236.

Fishman RA (1966). Blood-brain and CSF barriers to penicillin and related organic acids. *Arch Neurol*, *15*: 113-124.

Galinsky RE, Flaharty KK, Hoesterey BL and Anderson BD (1991). Probenecid enhances central nervous system uptake of 2',3'-dideoxyinosine by inhibiting cerebrospinal fluid efflux. *J Pharm Exp Therap*, *257*: 972-978.

Gerhart DZ, Enerson BE, Zhdankina OY, Leino RL and Drewes LR (1997). Expression of monocarboxylate transporter MCT1 by brain endothelium and glia in adult and suckling rats. *Am J Physiol*, *273*: E207-E213.

Gerhart DZ, Enerson BE, Zhdankina OY, Leino RL and Drewes LR (1998). Expression of the monocarboxylate transporter MCT2 by rat brain glia. *Glia*, *22*: 272-281.

Golden PL and Pardridge WM (2000). Brain microvascular P-glycoprotein and a revised model of multidrug resistance in brain. *Cell Mol Neurobiol*, *20*: 165-181.

Golden PL and Pollack GM (2003). Blood-brain Barrier Efflux Transport. *J Pharm Sci*, *92*: 1739-1753.

Gottesman MM and Pastan I (1993). Biochemistry of multidrug resistance mediated by the multidrug transporter. *Ann Rev Biochem, 62*: 385-427.

Gottesman MM, Pastan I and Ambudkar SV (1996). P-glycoprotein and multidrug resistance. *Curr Opin Genet Dev, 6*: 610-617.

Grundemann D, Babin-Ebell J, Martel F, Ording N, Schmidt A and Schomig E (1997). Primary structure and functional expression of the apical organic cation transporter from kidney epithelial LLC-PK_1 cells. *J Biol Chem, 272*: 10408-10413.

Gutmann H, Torok M, Fricker G, Huwyler J, Beglinger C and Drewe J (1999). Modulation of multidrug resistance protein expression in porcine brain capillary endothelial cells in vitro. *Drug Metab Disp, 27*: 937-941.

Halestrap AP and Price NT (1999). The proton-linked monocarboxylate transporter (MCT) family: structure, function and regulation. *Biochem J, 343*: 281-299.

Hedaya MA and Sawchuk RJ (1989). Effect of probenecid on the renal and nonrenal clearances of zidovudine and its distribution into cerebrospinal fluid in the rabbit. *J Pharm Sci, 78*: 716-722.

Hegmann EJ, Bauer HC and Kerbel RS (1992). Expression and functional activity of p-glycoprotein in cultured cerebral capillary endothelial cells. *Cancer Res, 52*: 6969-6975.

Hosoyamada M, Sekine T, Kanai Y and Endou H (1999). Molecular cloning and functional expression of a multispecific organic anion transporter from human kidney. *Am J Physiol, 276*: F122-F128.

Huai-Yun H, Secrest DT, Mark KS, Carney D, Brandquist C, Elmquist WF and Miller DW (1998). Expression of multidrug resistance-associated protein (MRP) in brain microvessel endothelial cells. *Biochem Biophys Res Comm, 243*: 816-820.

Inui K-I, Masuda S and Saito H (2000). Cellular and molecular aspects of drug transport in the kidney. *Kidney Int, 58*: 944-958.

Jackson VN, Price NT, Carpenter L and Halestrap AP (1997). Cloning of the monocarboxylate transporter isoform MCT2 from rat testis provides evidence that expression in tissues is species-specific and may involve post-transcriptional regulation. *Biochem J, 324*: 447-453.

Jette L, Pouliot JF, Murphy GF and Beliveau R (1995). Isoform 1 (mdr3) is the major form of p-glycoprotein expressed in mouse brain capillaries. *Biochem J, 305*: 761-766.

Jette L, Tetu B and Beliveau R (1993). High levels of p-glycoprotein detected in isolated brain capillaries. *Biochim Biophys Acta, 1150*: 147-154.

Katzman R, Graziani L, Kaplan R and Escriva A (1965). Exchange of cerebrospinal fluid potassium with blood and brain. Study in normal and Ouabain perfused cats. *Arch Neurol, 13(5)*: 513-524.

Kekuda R, Prasad PD, Wu X, Wang H, Fei YJ, Leibach FH and Ganapathy V (1998). Cloning and functional characterization of a potential-sensitive, polyspecific organic cation transporter (OCT3) most abundantly expressed in placenta. *J Biol Chem, 273*: 15971-15979.

Kim RB, Fromm MF, Wandel C, Leake B, Wood AJ, Roden DM and Wilkinson GR (1998). The drug transporter p-glycoprotein limits oral absorption and brain entry of HIV-1 protease inhibitors. *J Clin Invest, 101*: 289-294.

Koehler-Stec EM, Simpson IA, Vannucci SJ, Landschulz KT and Landschulz WH (1998). Monocarboxylate transporter expression in mouse brain. *Am J Physiol*, *275*: E516-E524.

Koepsell H (1998). Organic cation transporters in intestine, kidney, liver and brain. *Ann Rev Physiol*, *60*: 243-266.

Kool M, de Haas M, Scheffer GL, Scheper RJ, van Eijk MJ, Juijn JA, Baas F and Borst P (1997). Analysis of expression of cMOAT (MRP2), MRP3, MRP4, and MRP5, homologues of the multidrug resistance-associated protein gene (MRP1), in human cancer cell lines. *Cancer Res*, *57*: 3537-3547.

Kool M, van der Linden M, de Haas M, Baas F and Borst P (1999). Expression of human MRP6, a homologue of the multidrug resistance protein gene MRP1, in tissues and cancer cells. *Cancer Res*, *59*: 175-182.

Koziara JM, Lockman PR, Allen DD and Mumper RJ (2003). In situ blood-brain barrier transport of nanoparticles. *Pharm Res*, *20:* 1772-1778.

Koziara JM, Lockman PR, Allen DD and Mumper RJ (2004). Paclitaxel nanoparticels for the potential treatment of brain tumors. *J Contr Rel*, *99*: 259-269.

Kusuhara H, Sekine T, Utsunomiya-Tate N, Tsuda M, Kojima R, Cha SH, Sugiyama Y, Kanai Y and Endou H (1999). Molecular cloning and characterization of a new multispecific organic anion transporter from rat brain. *J Biol Chem*, 274: 13675-13680.Kusuhara H, Suzuki H, Naito M, Tsuruo T and Sugiyama Y (1998). Characterization of efflux transport of organic anions in a mouse brain capillary endothelial cell line. *J Pharm Exp Therap*, *285*: 1260-1265.

Kusuhara H, Suzuki H, Terasaki T, Kakee A, Lemaire M and Sugiyama Y (1997). P-glycoprotein mediates the efflux of quinidine across the blood-brain barrier. *J Pharm Exp Therap*, *283*: 574-580.

Lafreniere RG, Carrel L, and Willard HF (1994). A novel transmembrane transporter encoded by the XPCT gene in Xq13.2. *Hum Mol Genet*, *3*: 1133-1139.

Lechardeur D and Scherman D (1995). Functional expression of the p-glycoprotein mdr in primary cultures of bovine cerebral endothelial cells. *Cell Biol Toxciol*, *11*: 283-293.

Lee G and Bendayan R (2004). functional expression and localization of P-glycoprotein in the central nervous system: relevance to the pathogenesis and treatment of neurological disorders. *Pharm Res*, *21*: 1313-1330.

Leino RL, Gerhart DZ and Drewes LR (1999). Monocarboxylate transporter (MCT1) abundance in brains of suckling and adult rats: a quantitative electron microscopic immunogold study. *Dev Brain Res*, *113*: 47-54.

Lemaire M, Bruelisauer A, Guntz P and Sato H (1996). Dose-dependent brain penetration of SDZ PSC 833, a novel multidrug resistance-reversing cyclosporin, in rats. *Cancer Chemother Pharmacol*, *38*: 481-486.

Li D and Kerns EH (2003) Profiling drug-like properties in discovery research. *Curr Opin Chem Biol*, *7*: 402-408.

Lockman PR, Koziara J, Roder KE, Paulson J, Abbruscato TJ, Mumper RJ and Allen DD (2003). In vivo and in vitro assessment ofbaseline blood-brain barrier parameters in the presence of nanoparticles. *Pharm Res*, *20*: 705-713.

Lorico A, Rappa G, Finch RA, Yang D, Flavell RA and Sartorelli AC (1997). Disruption of the murine MRP (multidrug resistance protein) gene leads to increased sensitivity to etoposide (VP-16) and increased levels of glutathione. *Cancer Res*, *57*: 5238-5242.

Loscher W and Frey H-H (1982). Transport of GABA at the blood-CSF interface. *J Neurochem*, *38*: 1072-1079.

Madrid Y, Feigenbaum Langer L, Brem H and Langer R (1991). New Directions in the delivery of drugs and other substances to the central nervous system. *Adv Pharm*, *22*: 299-324.

Marzolini C, Paus E, Buclin T and Kim RB (2004). Polymorphisms in human MRD1 (P-glycoprotein): recent advances and clinical relevance. *Clin Pharmacol Ther 75*: 13-33.

Matsuoka Y, Okazaki M, Kitamura Y and Taniguchi T (1999). Developmental expression of P-glycoprotein (multidrug resistance gene product) in rat brain. *J Neurobiol*, *39*: 383-392.

Mayer U, Wagenaar E, Beijnen JH, Smit JW, Meijer DK, van Asperen J, Borst P and Schinkel AH (1996). Substantial excretion of digoxin via the intestinal mucosa and prevention of long-term digoxin accumulation in the brain by the mdr1a p-glycoprotein. *Br J Pharm*, *119*: 1038-1044.

Mayer U, Wagenaar E, Dorobek B, Beijnen JH, Borst P and Schinkel AH (1997). Full blockade of intestinal P-glycoprotein and extensive inhibition of blood-brain barrier p-glycoprotein by oral treatment of mice with PSC833. *J Clin Invest*, *100*: 2430-2436.

McAleer MA, Breen MA, White NL and Matthews N (1999). PABC11 (also known as MOAT-C and MRP5), a member of the ABC family of proteins, has anion transporter activity but does not confer multidrug resistance when overexpressed in human embryonic kidney 293 cells. *J Biol Chem*, *274*: 23541 –23548.

Mercier C, Masseguin C, Roux F, Gabrion J and Schermann J-M (2004). Expression of P-glycoprotein (ABCB1) and Mrp1 (ABCC1) in adult rat brain: focus on astrocytes. *Brain Res*, *1021*: 32-40.

Miller DS, Nobmann SN, Gutmann H, Toeroek M, Drewe J and Fricker G (2000). Xenobiotic transport across isolated brain microvessels studied by confocal microscopy. *Mol Pharm*, *58*: 1357-1367.

Murata M, Tamai I, Kato H, Nagata O and Tsuji A (1999). Efflux transport of a new quinolone antibacterial agent, HSR-903, across the blood-brain barrier. *J Pharm Exp Therap*, *290*: 51-57.

Nishino J, Suzuki H, Sugiyama D, Kitazawa T, Ito K, Hanano M and Sugiyama Y (1999). Transepithelial transport of organic anions across the choroid plexus: possible involvement of organic anion transporter and multidrug resistance-associated protein. *J Pharm Exp Therap*, *290*: 289-294.

Noe B, Hagenbuch B, Stieger B and Meier PJ (1997). Isolation of a multispecific organic anion and cardiac glycoside transporter from rat brain. *Proc Natl Acad Sci USA*, *94*: 10346-10350.

Ooie T, Terasaki T, Suzuki H and Sugiyama Y (1997). Kinetic evidence for active efflux transport across the blood-brain barrier of quinolone antibiotics. *J Pharm Exp Therap*, *283*: 293-304.

Pappenheimer JR, Heisey SR and Jordan EF (1961). Active transport of diodrast and phenolsulfonphthalein from cerebrospinal fluid to blood. *Am J Physiol, 200*: 1-10.

Pardridge WM, Golden PL, Kang YS and Bickel U (1997). Brain microvascular and astrocyte localization of P-glycoprotein. *J Neurochem, 68*: 1278-1285.

Pardridge WM (1998). CNS drug design based on principles of blood-brain barrier transport. *J Neurochem, 70*: 1781-1792.

Pardridge WM (2002). Drug and gene targeting to the brain with molecular Trojan horses. *Nat Rev Drug Discov, 1*:131-139.

Price NT, Jackson VN and Halestrap AP (1998). Cloning and sequencing of four new mammalian monocarboxylate transporter (MCT) homologues confirms the existence of a transporter family with an ancient past. *Biochem J, 329*: 321-328.

Prockop LD, Schanker LS and Brodie BB (1962). Passage of lipid insoluble substances from cerebrospinal fluid to blood. *J Pharm 135*: 266.

Rall DP and Sheldon W (1962). Transport of organic acid dyes by the isolated choroid plexus of the spiny dogfish S. acanthias. *Biochem Pharmacol,11*: 169-170.

Rao VV, Dahlheimer JL, Bardgett ME, Snyder AZ, Finch RA, Sartorelli AC and Piwnica-Worms D (1999). Choroid plexus epithelial expression of mdr1 P-glycoprotein and multidrug resistance-associated protein contributes to the blood-cerebrospinal-fluid drug-permeability barrier. *Proc Natl Acad Sci USA, 96*: 3900-3905.

Reed DJ, Woodbury DM, Jacobs L and Squires R (1965). Factors affecting distribution of iodide in brain and cerebrospinal fluid. *Am J Physiol, 209*: 757-765.

Regina A, Koman A, Piciotti M, El Hafny B, Center MS, Bergmann R, Couraud PO and Roux F (1998). Mrp1 multidrug resistance-associated protein and p-glycoprotein expression in rat brain microvessel endothelial cells. *J Neurochem, 71*: 705-715.

Rousselle C, Clair P, Lefauconnier JM, Kaczorek M, Scherrmann JM and Temsamani J (2000). New advances in the transport of doxorubicin through the blood-brain barrier by a peptide vector-mediated strategy. *Mol Pharmacol, 57*: 679-686.

Schanker LS, Prockop LD, Schou J, Sisodia P (1962). Rapid efflux of some quaternary ammonium compounds from cerebrospinal fluid. *Life Sci, 10*:515-521.

Schinkel AH, Smit JJ, van Tellingen O, Beijnen JH, Wagenaar E, van Deemter L, Mol CA, van der Valk MA, Robanus-Maandag EC, te Riele HP, Berns AJM and Borst P (1994). Disruption of the mouse mdr1a p-glycoprotein gene leads to a deficiency in the blood-brain barrier and to increased sensitivity to drugs. *Cell, 77*: 491-502.

Schinkel AH, Wagenaar E, Mol CA and van Deemter L (1996). P-glycoprotein in the blood-brain barrier of mice influences the brain penetration and pharmacological activity of many drugs. *J Clin Invest, 97*: 2517-2524.

Schinkel AH, Wagenaar E, van Deemter L, Mol CA and Borst P(1995). Absence of the mdr1a p-glycoprotein in mice affects tissue distribution and pharmacokinetics of dexamethasone, digoxin, and cyclosporin A. *J Clin Invest, 96*: 1698-1705.

Schinkel AH (1999). P-glycoprotein, a gatekeeper in the blood-brain barrier. *Adv Drug Deliv Rev, 36*: 179-194.

Schumacher U and Mollgård K (1997). The multidrug-resistance P-glycoprotein (P-gp, mdr1) is an early marker of blood-brain barrier development in the microvessels of the developing human brain. *Histochem Cell Biol, 108*: 179-182.

Seetharaman S, Barrand MA, Maskell L and Scheper RJ (1998). Multidrug resistance-related transport proteins in isolated human brain microvessels and in cells cultured from these isolates. *J Neurochem, 70*: 1151-1159.

Sekine T, Cha SH and Endou H (2000). The multispecific organic anion transporter (OAT) family. *Eur J Physiol, 440*: 337-350.

Siddiqui A, Kerb R, Weale ME, Brinkmann U, Smith A, Goldstein DB, Wood NW and Sisodiya SM (2003) Association of multidrug resistance in epilepsy with a polymorphism in the drug-transporter gene ABCB1. *New Engl J Med, 348*: 1442-1448.

Sonneveld P and Wiemer E (1997). Inhibitors of multidrug resistance. *Curr Opin Oncol 9*: 543-548.

Spector R and Goetzl EJ (1986). Leukotriene C4 transport and metabolism in the central nervous system. *J Neurochem, 46*: 1308-1312.

Spector R and Lorenzo AV (1974). The effects of salicylate and probenecid of the cerebrospinal fluid transport of penicillin, aminosalicylic acid and iodide. *J Pharm Exp Therap, 188:* 55-65.

Stewart PA, Beliveau R and Rogers KA (1996). Cellular localization of P-glycoprotein in brain versus gonadal capillaries. *J Histochem Cytochem, 44*: 679-685.

Stride BD, Valdimarsson G, Gerlach JH, Wilson GM, Cole SP, Deeley RG (1996). Structure and expression of the messenger RNA encoding the murine multidrug resistance protein, an ATP-binding cassette transporter. *Mol Pharmacol, 49:* 962-971.

Sugawara I, Hamada H, Tsuruo T and Mori S (1990). Specialized localization of P-glycoprotein recognized by MRK16 monoclonal antibody in endothelial cells of the brain and spinal cord. *Jpn J Cancer Res, 81*: 727-730.

Suzuki H, Sawada Y, Sugiyama Y, Iga T, Hanano M and Spector R (1989). Transport of imipenem, a novel carbarpenem antibiotic, in the rat central nervous system. *J Pharm Exp Therap, 250*: 979-984.

Takanaga H, Tamai I and Inaba S (1995). cDNA cloning and functional characterization of rat intestinal monocarboxylate transporter. *Bioch Biophys Res Comm, 217*: 370-377.

Takasawa K, Terasaki T, Suzuki H and Sugiyama Y (1997) In vivo evidence for carrier-mediated efflux transport of 3'-azido-3'-deoxythymidine and 2',3'-dideoxyinosine across the blood-brain barrier via a probenecid-sensitive transport system. *J Pharm Exp* Therap, *281*: 369-375.

Tamai I, Ohashi R, Nezu JI, Sai Y, Kobayashi D, Oku A, Shimane M AND Tsuji A (2000). Molecular and functional characterization of organic cation/carnitine transporter family in mice. *J Biol Chem, 275*:40064-40072.

Tamai I, Ohashi R, Nezu J, Yabuuchi H, Oku A, Shimane M, Sai Y and Tsuji A (1998). Molecular and functional identification of sodium ion-dependent, high affinity human carnitine transporter OCTN2. *J Biol Chem, 273*: 20378-20382.

Tamai I, Yabuuchi H, Nezu J, Sai Y, Oku A, Shimane M and Tsuji A (1997). Cloning and characterization of a novel human pH-dependent organic cation transporter, OCTN1. *FEBS Lett, 419*: 107-111.

Tatsuta T, Naito M, Oh-hara T, Sugawara I and Tsuruo T (1992). Functional involvement of p-glycoprotein in blood-brain barrier. *J Biol Chem, 267*: 20383-20391.

Taylor EM (2002). The impact of efflux transporters in the brain on the development of drugs for CNS disorders. *Clin Pharmacokinet*, *41*: 81-92.

Thiebaut F, Tsuruo T, Hamada H, Gottesman MM, Pastan I and Willingham MC (1989). Immunohistochemical localization in normal tissues of different epitopes in the multidrug transport protein P170: evidence for localization in brain capillaries and crossreactivity of one antibody with a muscle protein. *J Histochem Cytochem*, *37:* 159-164.

Tishler DM, Weinberg KI, Hinton DR, Barbaro N, Annett GM and Raffel C (1995). MDR1 gene expression in brain of patients with medically intractable epilepsy. *Epilepsia*, *36*: 1-6.

Toth K, Vaughan MM, Peress NS, Slocum HK and Rustum YM (1996). MDR1 P-glycoprotein is expressed by endothelial cells of newly formed capillaries in human gliomas but is not expressed in the neovasculature of other primary tumors. *Am J Pathol*, *149*: 853-858.

Tsuji A, Terasaki T, Takabatake Y, Tenda Y, Tamai I, Yamashima T, Moritani S, Tsuruo T and Yamashita J (1992). P-glycoprotein as the drug efflux pump in primary cultured bovine brain capillary endothelial cells. *Life Sci*, *51*: 1427-1437.

Volm M (1998). Multidrug resistance and its reversal. *Anticancer Res*, *18:* 2905-2918.

Walters HC, Craddock AL, Fusegawa H, Willingham MC and Dawson PA (2000). Expression, transport properties, and chromosomal location of organic anion transporter subtype 3. *Am J Physiol*, *279*: G1188-1200.

Wang Y and Sawchuk RJ (1995). Zidovudine transport in the rabbit brain during intravenous and intracerebroventricular infusion. *J Pharm Sci*, *84*: 871-876.

Wang Y, Wei Y and Sawchuk RJ (1997). Zidovudine transport within the rabbit brain during intracerebroventricular administration and the effect of probenecid. *J Pharm Sci*, *86*: 1484-1493.

Warren KE, Patel MC, McCully CM, Montuenga LM and Balis FM (2000). Effect of P-glycoprotein modulation with cyclosporin on cerebrospinal fluid penetration of doxorubicin in non-human primates. *Cancer Chemotherap Pharmacol*, *45*: 207-212.

Waterhouse RN (2003). Determination of lipophilicity and its use as a predictor of blood-brain barrier penetration of molecular imaging agents. *Mol Imaging Biol*, *5*:376-89.

Weiner IM and Mudge GH (1964). Renal tubular mechanisms for excretion of organic acids and bases. *Am J Med*, *36*: 743-762.

Welch K (1962a). Active transport of iodide by choroid plexus of the rabbit in vitro. *Am J Physiology*, *202*:757-760.

Welch K (1962b). Concentration of thiocyanate by the chroroid plexus of the rabbit in vitro. *Proc Soc Exp Biol NY*, *109*: 953-954.

Wijnholds J, deLange EC, Scheffer GL, van den Berg DJ, Mol CA, van der Valk M, Schinkel AH, Scheper RJ, Breimer DD and Borst P (2000). Multidrug resistance protein 1 protects the choroid plexus epithelium and contributes to the blood-cerebrospinal fluid barrier. *J Clin Invest*, *105*: 279-285.

Wijnholds J, Evers R, van Leusden MR, Mol CA, Zaman GJ, Mayer U, Beijnen JH, van der Valk M, Krimpenfort P and Borst P (1997). Increased sensitivity to anticancer drugs and decreased inflammatory response in mice lacking the multidrug resistance-associated protein. *Nature Med*, *3*: 1275-1279.

Wijnholds J, Scheffer GL, van der Valk M, van der Valk P, Beijnen JH, Scheper RJ and Borst P (1998). Multidrug resistance protein 1 protects the oropharyngeal mucosal layer and the testicular tubules against drug-induced damage. *J Exp Med*, *188*: 797-808.

Wong SL, van Belle K and Sawchuk RJ (1993). Distributional transport kinetics of zidovudine between plasma and brain extracellular fluid/cerebrospinal fluid in the rabbit: investigation of the inhibitory effect of probenecid utilizing microdialysis. *J Pharm Exp Therap*, *264*: 899-909.

Wu X, George RL, Huang W, Wang H, Conway SJ, Leibach FH, Ganapathy V (2000). Structural and functional characteristics and tissue distribution pattern of rat OCTN1, an organic cation transporter, cloned from placenta. *Biochim Biophys Acta*, *1466*: 315-327.

Wu X, Huang W, Prasad PD, Seth P, Rajan DP, Leibach FH, Chen J, Conway SJ and Ganapathy V (1999). Functional characteristics and tissue distribution pattern of organic cation transporter 2 (OCTN2), an organic cation/carnitine transporter. *J Pharm Exp Therap*, *290*: 1482-1492.

Wu X, Kekuda R, Huang W, Fei YJ, Leibach FH, Chen J, Conway SJ and Ganapathy V (1998). Identity of the organic cation transporter OCT3 as the extraneuronal monoamine transporter (uptake$_2$) and evidence for the expression of the transporter in the brain. *J Biol Chem*, *273*: 32776-32786.

Yan R and Taylor EM (2002). AIT-082 is transported out of brain by a saturable mechanism: possible involvement of multidrug resistance and monocarboxylic acid transporters. *Drug Disp Metab*, *30*:513-8.

Zaman GJ, Versantvoort CH, Smit JJ, Eijdems EW, de Haas M, Smith AJ, Broxterman HJ, Mulder NH, de Vries EG, Baas F and Borst P (1993). Analysis of the expression of MRP, the gene for a new putative transmembrane drug transporter, in human multidrug resistant lung cancer cell lines. *Cancer Res*, *53*: 1747-1750.

Zhang L, Dresser MJ, Gray AT, Yost SC, Terashita S and Giacomini KM (1997). Cloning and functional expression of a human liver organic cation transporter. *Mol Pharm*, *51*: 913-921.

Zhang Y, Han H, Elmquist WF and Miller DW (2000). Expression of various multidrug resistance-associated protein (MRP) homologues in brain microvessel endothelial cells. *Brain Res*, *876*: 148-153.

Zimprich F, Sunder-Plassmann R, Stogmann E, Gleiss A, Dal-Bianco A, Zimprich A, Plumer S, Baumgartner C and Mannhalter C (2004). Association of an ABCB1 gene haplotype with pharmacoresistance in temporal epilepsy. *Neurol*, *63*: 1087-1089.

In: Focus on Brain Research
Editor: Casey J. Resch, pp. 107-121
ISBN 978-1-60021-395-3

Chapter IV

THE ASSESSMENT OF FUNCTIONAL BRAIN LATERALIZATION: METHODOLOGICAL ASPECTS

***Andreas Jansen*[1] *and Stefan Knecht*[1,2]**
[1]Department of Neurology
[2]IZKF Münster, University of Münster, Germany

ABSTRACT

The brain areas subserving cognitive functions frequently show an unequal involvement across hemispheres. The left hemisphere's superiority for language and the right hemisphere's superiority for tasks like attentional monitoring are the most obvious of these functional lateralizations. The investigation of hemispheric lateralization is not only of theoretical interest, but also relevant in the clinical context. For instance, the degree of language lateralization is a marker for the susceptibility to unilateral brain lesions, and a lack of hemispheric control due to decreased lateralization is discussed as one of the mechanisms contributing to various neuropsychiatric disorders, like autism, attention deficit hyperactivity disorder or schizophrenia.

This chapter is divided into two sections. In the first part, we present a short overview on the functional lateralization of cognitive brain functions and discuss applications and limitations of laterality research in various clinical and non-clinical domains. In the second section we outline some methodological aspects that have to be addressed in fMRI laterality studies to deliver meaningful results. Special emphasis is given to the question how - in terms of reliability and information content - the lateralization of a cognitive function can be characterized.

I. HEMISPHERIC LATERALIZATION OF BRAIN FUNCTIONS

The hemispheres of the brain, which by first glance appear to be identical and anatomically symmetrical, are functionally fundamentally different. Today, functional asymmetry is known as a basic principle of human brain organisation. Whether it is recognition of faces (Rossion *et al.* 2003), processing of emotion (Wildgruber *et al.* 2005), spatial orientation (Vogel *et al.* 2003) or identification of subtle rhythm-difference in music (Vuust *et al.* 2005) – most cognitive functions are now understood to be asymmetrically implemented in the brain. To understand the functioning of the human brain, we have to be able to understand the basic principles it follows. Consequently, we also have to be able to account for the functional asymmetry it shows in many of its functions.

One of the best-established lateralized functions of the brain is human language. First discoveries on the lateralization of language were won by clinical observation around the middle of the 19th century. Patients with lesions of certain areas of the left hemisphere showed language disorders, while patients with lesions of the corresponding areas of the right hemisphere less often showed a similar loss of language (Broca 1861a; Broca 1861b). This led to the conclusion that core functions of language are lateralized to the left hemisphere of the brain. Paul Broca was one of the first to formulate this conclusion. Already in 1861 he wrote (Broca 1861a):

> «Tout permet donc de croire que, dans le cas actuel, la lésion du lobe frontal a été la cause de la perte de la parole» [1]

A multitude of studies has investigated the functions of the individual hemispheres, distinguishing between tasks which can or cannot be computed by the left or right part of the brain. The reason why the anatomically similar hemispheres differ so strongly in their functional specialization however remains an open question. Equally, the physical mechanism constituting these differences remains a puzzle to be solved.

In spite of this we might be able to answer the question why language processes are lateralized at all. Language is one of the most complex cognitive functions. In particular, temporal coordination of neuronal excitation is of importance for language tasks (Friederici and Alter 2004). Further it is critical to quickly convey neuronal information. Since neuronal distances within one hemisphere are shorter than distances between two hemispheres, information should be conveyed more quickly when language functions are localized within one hemisphere and not spread out between both. Thus, from an evolutionary point of view lateralized language would offer an advantage in selection. This account provides a tentative explanation why language is lateralized, but it cannot offer any explanation why this function is lateralized to the left (Knecht *et al.* 2001). Possibly, neurons of the left hemisphere offered a better environment for the development of language (Garcia *et al.* 2004). Evolutionary, spoken language might trace back to far more ancient means of communication like gestures and mimics (Corballis 1992; Rizzolatti and Arbib 1998; Floel *et al.* 2003). Neurons, which were implicated in the use of the dominant hand, may also have been used for the neuronal

[1] Everything leads to believe that, in this case, the frontal lobe lesion was the cause of the loss of speech.

connectivity of language. According to alternative theories language developed in the left hemisphere because the right hemisphere was already involved in emotions and tasks of orientation, both of which are early evolutionary functions highly important for survival. Consequently, the left hemisphere could offer more space for the implementation of a new cognitive function.

Determination of Hemispheric Dominance of Cognitive Brain Functions

Since the 19th century it has been a major neuroscientific goal to associate distinct cortical regions with specific cognitive functions. By correlating microstructurally defined brain areas with lesion sites found in patients with specific neuropsychological symptoms, cognitive brain functions could be assigned to circumscribed brain areas in one hemisphere. This approach however has severe limitations. Due to the compensatory mechanism of the brain the relation between specific lesions and neuropsychological measurable deterioration of cognitive abilities is sometimes difficult to assess, leading to major problems of interpretation.

Modern imaging techniques have promised an alternative approach to the "lesion model". However, for a long time the reliable assessment of hemispheric dominance could only be achieved by invasive methods, for instance by injection of amobarbital (IAP, intracarotid amobarbital procedure). By selectively inactivating one hemisphere, the functional role of the other hemisphere for a particular cognitive function can be determined (Wada and Rasmussen 1960). Since the technique carries the risk of fatal complications, it is therefore only administered to neurosurgical candidates. These subjects, however, often suffered brain damage early in life, which may have led to substantial reorganization of brain functions. This makes it difficult to generalize to the normal functions of particular brain areas.

With the advent of non-invasive imaging techniques about 20 years ago, it became possible to non-invasively investigate the lateralization of cognitive brain functions in the healthy population. Hemispheric dominance of cognitive functions can be determined with different functional imaging modalities such as functional magnetic resonance imaging (fMRI) (Heeger and Ress 2002), functional transcranial Doppler sonography (fTCD) (Deppe *et al.* 2004), positron emission tomography (PET) (Votaw 1995), magnetoencephalography (MEG) (Lutkenhoner 2003) and near infrared optical topography (OT) (Kennan *et al.* 2002). Several pioneering studies successfully estimated language dominance in agreement with the Wada-test, which is considered as the gold standard for determining language lateralization (Binder *et al.* 1996; Knecht *et al.* 1998; Knake *et al.* 2003; Hirata *et al.* 2004). While the Wada-test directly tests for hemispheric dominance by selectively inhibiting one hemisphere, less invasive functional imaging techniques measure the brain activity elicited by sensory, motor or cognitive tasks. Some techniques (such as MEG) directly measure the neural activity, other techniques are sensitive for changes in cerebral blood flow differences (like PET) or blood oxygen-level dependent (BOLD) changes (like fMRI).

In the last decade, a considerable body of data has been collected on the variability of the language system, largely by means of fMRI and fTCD. It has been established that language can be produced and perceived by predominantly the left hemisphere, predominantly the right one, or by both hemispheres, unrelated to gender or linguistic performance (Frost *et al.* 1999; Pujol *et al.* 1999; Knecht *et al.* 2001). About 10% of humans have been found to have an atypical, i.e. bilateral or right-hemisphere representation for language. The incidence of right-hemisphere language dominance increases linearly with the degree of left-handedness, from 4% in strong right-handers to 15% in ambidextrous individuals and 27% in strong left-handers (Knecht *et al.* 2000). These results have been taken as evidence that the relationship between handedness and language dominance is not an artefact of cerebral pathology but a natural phenomenon.

Applications and Limitations of Laterality Research

The exploration of hemispheric asymmetry is considered to be important, in the clinical context as well as in the theoretical neurosciences. This can be demonstrated by a simple literature search. According to *PubMed* (www.pubmed.org), one of the most important literature databases for medicine and neuroscience, more than 3000 articles were published since 1990 that referred to the search criteria[2] "hemispheric asymmetry", "lateralization", or "hemispheric dominance".

The degree of language lateralization, for instance, is taken as a marker for the susceptibility to unilateral brain lesions (Knecht *et al.* 2002). The determination of language lateralization is important for the presurgical evaluation of patients with medically intractable epilepsy, because the risk of postsurgical language and memory deficits is related directly to the presurgical lateralization of language functions in the brain (Bell *et al.* 2000; Sabsevitz *et al.* 2003). The lateralization of cognitive functions has also become a subject of interest in psychiatry. Various disorders as autism (Hier *et al.* 1979), attention deficit hyperactivity disorder (ADHD) (Mostofsky *et al.* 2002), and schizophrenia (Bertolino *et al.* 2004) have been related to decreased laterality. Disturbances in the development of lateralization have been also linked to dyslexia (Heim and Keil 2004). Another field of laterality research is the study of genetic cohorts. By providing information on the differences between individuals in brain activation, functional imaging can be applied in the validation of theoretical models about a genetic basis of the lateralization of higher cognitive functions (McManus 1985; Annett and Alexander 1996). Hemispheric dominance has also been embraced by the popular sciences and thus by a more general audience (Pease and Pease 1998), although here the behavioural consequences of functional brain asymmetry are often clouded by oversimplifications and myths, for instance - to give a popular example - the presumed female inability to back into a parking space that has been attributed to a reduced laterality of cognitive functions.

Most classical approaches towards brain-function relationships are based on correlational approaches. A specific brain function is attributed to specific brain areas, expressing a direct causal role of structure for function (Stephan 2004). However, cortical areas do not operate in isolation but are connected to many other brain regions. Therefore, the functional behavior of one brain area can not be explained from its local microstructure alone, but inputs from other brain areas have to be taken into consideration as well. Therefore the correlative approach, which has beyond doubt generated a huge amount of information about which regions are elements of the neural system governing certain cognitive processes, has severe limitations. The diversity of findings about the activation of specific brain regions in many cognitive tasks speaks against a one-to-one relation between specific cortical areas and specific brain functions. For a deeper understanding of the brain, functional neuroimaging must therefore go beyond simply demonstrating significant correlations between signal changes in certain brain regions and task components of interest. If not, all results are restricted to the description of regional functional specialization. In many neuroimaging studies, a neuronal system is simply described by a listing of brain areas activated during a certain task. Experiments are performed in an explanatory fashion, and results are explained post-hoc by some sort of story (Kosslyn 1999). In some aspects, this approach resembles *phrenology* (Box A). Newer approaches therefore propose models that reflect the connectional structure of the brain and emphasize context-depending couplings between the elements of neuronal systems (Friston *et al.* 2003).

Also the classical laterality research addresses the functional asymmetry of the human brain often simply by the description of a cognitive function in terms of its inter-hemispheric distribution, disregarding the coupling of different brain areas participating in a cognitive task and thus the connectional structure of the brain. For example, it has been shown that diseases such as schizophrenia are associated with lateralization abnormalities (Weiss *et al.* 2004), and alterations of mood and cognition observed during the menstrual cycle of women are linked to laterality changes in the language domain (Fernandez *et al.* 2003). This is definitely important as a first step towards more thorough understanding of both phenomena, however the interplay of brain areas must not be neglected. Classical laterality research has beyond doubt contributed to an understanding of the human (and also non-human) brain organization, but it now has also to adopt modern approaches and must not stop at a purely descriptive level.

Box A:

Phrenology was a science of character divination, popular at the end of the 18th and the beginning of 19th century. Human brain organization was described under the premise that the mind is composed of multiple distinct, innate "faculties" (such as memory, reason, intellect). Each faculty was assigned a separate seat in the brain. It was assumed that the skull takes its shape from the brain, and therefore the surface of the skull was used to determine which characteristics were most prominent. Even modern functional imaging studies often describe mental processes simply by observing which brain areas are activated during a cognitive task, but neglect the connectional structure of the brain. Although this approach has generated much information about the functioning of the brain, it falls short of explaining its underlying principles. The same applies to classical laterality research. It represents an important contribution in the attempt to understand the principles underlying brain organization, but has now to go beyond a purely descriptive level.

[2] The query syntax used was: "hemispheric asymmetry" OR lateralization OR lateralisation OR "hemispheric dominance"

II. The Quantification of Hemispheric Dominance fMRI Laterality Studies

One of the first steps in laterality research is always the determination of hemispheric dominance during a cognitive task. However, evaluating the hemispheric dominance in functional neuroimaging, especially in fMRI, is not simple. In the following section we outline several issues that have to be addressed for laterality studies to deliver meaningful results. These points are illustrated with data from studies investigating the neural correlates of atypical dominance for language and spatial attention.

In the neurosciences, the relative contribution of the hemispheres during a task is expressed by a lateralization index (LI), describing both *hemispheric dominance* (i.e. left or right hemispheric dominance) and the *degree of lateralization* (i.e. how strong left or right). However, the optimal method for evaluating the lateralization of a cognitive function is controversial. To be a useful marker of inter-hemispheric brain organization, a suitable approach has to meet the criteria of robustness reproducibility and information content (Box B).

Box B. Criteria for laterality measures

Robustness: Laterality calculations depend on decisions concerning the effect size measure and a precise region-of-interest (ROI). Both in turn depend on a number of external parameters. These choices are partially constrained by the questions being addressed, but additional arbitrary decisions remain. A suitable method for evaluating the lateralization of a cognitive function has to be robust, that is, the LI must not vary strongly with external parameters.

Reproducibility: FMRI measurements are not only variable across subjects, but also within subjects when tested for a second time. In the same subject, magnitude and extent of brain activation can change considerably. In the assessment of the clinical and theoretical usefulness of lateralization measures, intra-subject reliability is thus an important issue. Therefore, a suitable method for evaluating the lateralization of a cognitive function has to be reproducible, that is, yield similar results in repeated measurements.

Capability of inter-subject comparisons: A LI might be robust and reproducible, but to be a useful measure it must be capable to discriminate between subjects. For instance, if a LI describes the language lateralization always by the same fixed value for all subjects, it does not deliver any information about differences in subjects' brain organization although it might fulfil the criteria of robustness and reproducibility.

In most studies, a lateralization index is calculated by the formula

$$LI = \frac{A_L - A_R}{A_L + A_R},$$

where A_L and A_R refer to values of fMRI-measured activity for equal ROIs within the left (L) and right (R) hemisphere. A positive value of LI represents left-hemisphere dominance, a negative value right-hemisphere dominance. The intermediate values reflect varying degrees of laterality.

In the last decade, several approaches have been established to describe the brain activity $A_{L/R}$ and to choose a suitable ROI. All approaches are dependent on external parameters (e.g., the statistical threshold at which voxels are declared as "active"). These choices have to be made explicit, and it has to be shown that the effects of these decisions do not influence the overall findings. Only then can functional neuroimaging provide meaningful laterality results.

(a) Choice of Activity Measure $A_{L/R}$

To quantify brain activity, typically two approaches have been taken in fMRI studies. The first is to measure brain activity based on the *extent* of the activated brain region, i.e., the number of active voxels. The second is to measure the *magnitude* of the fMRI signal change within a ROI. In the following, a LI based on the extent of activated brain volume is labelled LI_e, a LI based on the magnitude of signal change LI_m.

Since the beginning of fMRI, the most popular approach in practice has been to calculate A_L and A_R by the volume of significantly activated brain above a given statistical threshold p in a given ROI. A drawback of this measure is its dependence on the statistical threshold p. At low thresholds (high p-values), the occurrence of false positive voxels in both hemispheres typically leads to a bilateral activation pattern, moving LI towards zero. At high thresholds (low p-values) the number of active voxels decreases, false positive voxels as well as "truly" activated voxels are excluded. LI_e approaches +/- 1. Thus, LI_e can express almost the complete continuum between bilaterality and fully lateralized brain activity (Figure 1). In some cases, a change of the statistical threshold can even reverse the sign of laterality and consequently hemispheric dominance (Figure 2).

Because of the strong dependence of LI_e on the statistical threshold p, its absolute value and thus the degree of lateralization is as arbitrary as the choice of p. LI_e based on fixed statistical thresholds has in no way the capability of inter-subject comparisons. Due to the high inter-subject variability in activation levels, two subjects may have a very similar activation pattern and thus a similar LI, but at completely different statistical thresholds (Figure 3). Therefore, several studies support the view that individually adapted thresholds are a reasonable means of analysing single subject data to obtain comparable results for several individuals and increase intra-subject reliability (Specht *et al.* 2003).

There have been several attempts to deal with the dependence of LI_e on arbitrary statistical thresholds as well as to cope with the high inter-individual variability in general activation levels. In plotting LI_e as a function of the statistical threshold, Binder et al. observed that, in most cases, the LI_e-curve reaches a plateau for a broader range of thresholds (Binder *et al.* 1996) (cf. Figure 1 right). Calculating LI_e for p-values within such a plateau makes the value less dependent on external parameters, because it allows for some variation in statistical thresholding without changes in LI_e. However, such a plateau often is not found (Nagata *et al.* 2001). Ramsey et al. used a combined task analysis (CTA) approach (Ramsey *et al.* 2001).

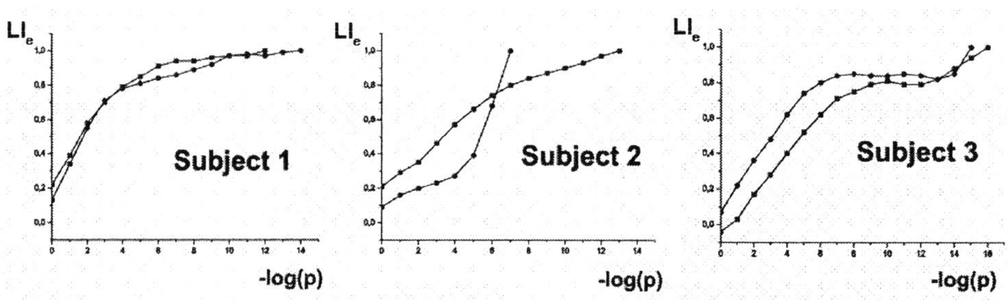

Figure 1. Lateralization index LIe as a function of the statistical threshold p for representative subjects. LIe is calculated by the volume of significantly activated brain volume during word generation in the left and right cerebral hemisphere, respectively. Depending on the threshold, LIe expresses almost the complete continuum between bilaterality and fully lateralized brain activity. It can be seen that the LIe(p) curve is reproducible in some subjects, but not in others. (Figure modified from (Jansen et al. 2005)).

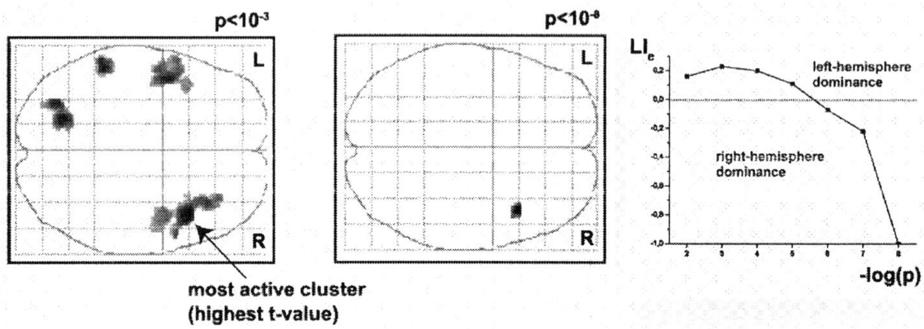

Figure 2. Activation patterns for a selected subject during the Landmark task, a task related to spatial attention, presented as through-projection onto a transversal representation of standard MNI space. LIe is calculated by the volume of significantly activated brain volume in the left and right cerebral hemisphere, respectively. The majority of the activation centres is located in the left hemisphere (left), the most significant activated cluster however is in the right hemisphere (middle). Thus, at low statistical thresholds, the subject has more activated brain volume in the left hemisphere and is therefore classified as left-hemisphere dominant. At high thresholds, only the most significant activated cluster is still declared as active, now indicating right-hemisphere dominance (right). For this subject, not only the degree of lateralization but also hemispheric dominance is a function of the statistical threshold. Abbreviations used: L - left, R - right. (Figure modified from (Jansen et al. 2005)).

By combining several language tasks, this approach is geared toward identifying brain regions that are involved in generic language functions rather than regions that are involved in functions that are specific to a single task. The authors showed that the CTA approach yielded laterality indices that are more consistent across statistical thresholds than individual task analysis. Knecht et al. set the level of significance in each subject so that it corresponded to a fixed total number of activated voxels (Knecht *et al.* 2003).

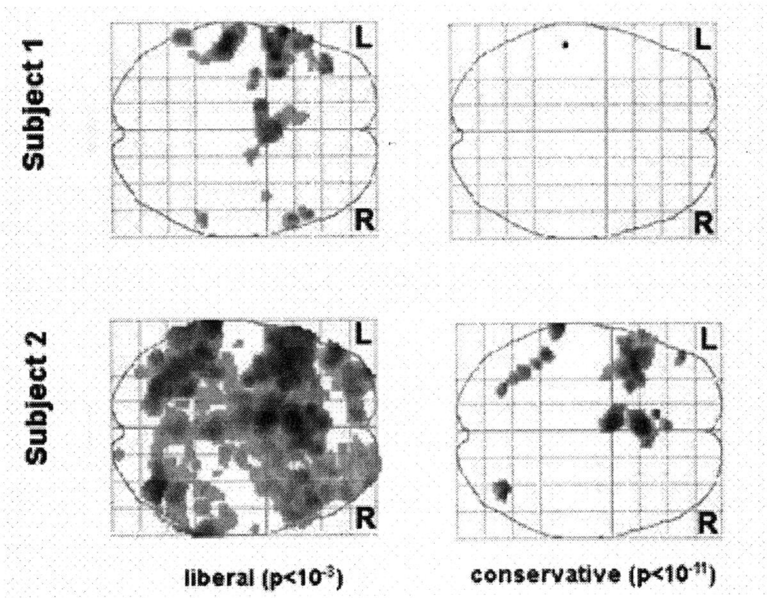

Figure 3. Activated brain regions associated with a word generation task presented as through-projection onto a transversal representation of stereotactic space. Both subjects show similar activation pattern, but at different statistical thresholds. At a fixed threshold, LIe for at least one of the subjects would be calculated for a unreasonable activation pattern. Given the high inter-subject variability in activation levels, for an inter-subject comparison no generally predefined threshold should be applied. Instead, statistical thresholds have to be individually adjusted. Abbreviations used L - left, R - right. (Figure modified from (Jansen et al. 2005)).

On the assumption that the subjects should engage the same amount of brain tissue - they all performed the same task at more or less the same level - they postulated that the extent of activated brain volume is similar for all subjects. The number of activated voxels were chosen "reasonably", that is, in accordance with the results of lesion studies as well as with other functional imaging studies.

Another approach is to base $A_{L/R}$ on the magnitude signal intensity change between two conditions. This approach is considered to be a more robust measure of brain activation than LI_e, because in principle no choice of threshold is involved in calculating these indices (Cohen and DuBois 1999; Adcock *et al.* 2003; Jansen *et al.* 2004). However, although LI_m is independent of statistical thresholds, this measure has its own set of problems. Its robustness is strongly depending on the chosen ROI. In the majority of cases, a LI is calculated with normalized brain images using the same ROI for all subjects. Because the spatial normalization procedure introduces spatial imprecision of activation, the activated clusters of different subjects might be displaced relative to each other. Therefore, ROIs do not have to be chosen too small, because otherwise the activation of interest might be outside the ROI for some subjects. Large ROIs however reduce the robustness of LI_m. Implicit assumption for the calculation of LI_m is that one deals with activations (i.e., stronger activation in the activation task than in the control task). If there is a deactivation in the ROI of the subdominant hemisphere (i.e., stronger activation in the control task than in the activation task), LI_m can become greater than +/-1, i.e. loses a meaningful metric. If the deactivation in the subdominant hemisphere is even greater than the activation in the dominant hemisphere, the sign of the lateralization index is reversed, i.e., LI_m denotes the subdominant hemisphere as dominant (Figure 4).

A solution to this problem has been proposed by Fernandez et al. (Fernandez *et al.* 2001). For the calculation of the effect size measure $A_{L/R}$ the authors did not account for all voxels in a chosen ROI, but only those with the strongest activation. In a first step, they calculated for each subject and each ROI a mean maximum activation value defined as the mean of those 5% of voxels showing the highest level of activation in the respective ROI. In a second step, they set the threshold for inclusion in the calculation of the LI at 50% of this mean maximum activation value. Only signal intensity changes of those voxels that exceeded the threshold were used to calculate LI_m. This had the advantage that weakly activated voxels, possibly showing a deactivation, are excluded. Applying this approach to the example shown in Figure 4, the LI based on magnitude of signal change shifted from LI_m=7.3 (all voxels in ROI, indicating hemispheric dominance wrong) to LI_m=-0.61.

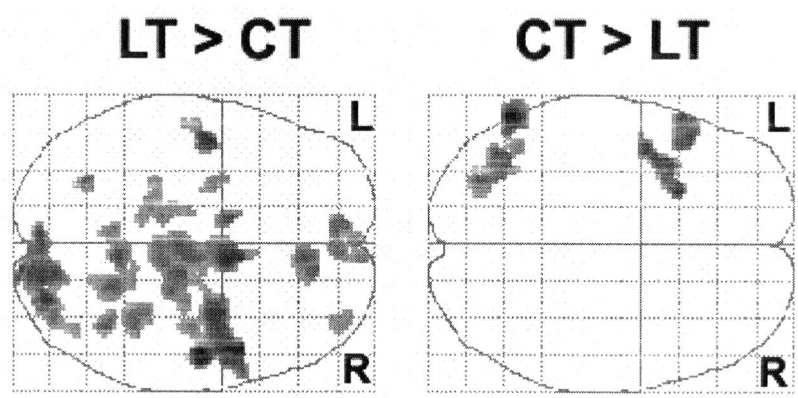

Figure 4. Activation pattern for a selected subject during the Landmark task, a task associated with spatial attention, presented as through-projection onto a transversal representation of MNI space. Different LIs are calculated for a ROI in the frontal cortex. Based on the extent of activation, the subject is clearly right-hemisphere dominant (LIe(frontal) = -0.9, p<10-3) with little activation in the left hemisphere (left, LT>CT, Landmark task > Control task). If the laterality index is calculated by the magnitude of the mean signal intensity change, the subject displays left hemispheric dominance for the frontal activation (LIm(frontal) = 7.3). A more detailed analysis of its activation pattern reveals, that this does not describe the "true" hemispheric dominance. Implicit assumption was that one deals with activation in both the left and the right ROI. As one can see in the reverse contrast (right, CT > LT, Control task > Landmark task), there is a strong "deactivation" in the left ROI, i.e., stronger activation in the control task than in the actual activation task. The deactivation in the subdominant hemisphere is greater than the activation in the dominant hemisphere. This leads to a reversal of laterality. Abbreviations used: L - left, R - right. (Figure modified from (Jansen et al. 2005)).

(b) Choice of a ROI

Cognitive functions are subserved by large-scale neurocognitive networks. Regions within this network show different extents of lateralization. By averaging over all activations, the LI is unlikely to reflect a systematic pattern of lateralization in the key participating brain regions. Thus, another important issue in laterality studies is how to determine a precise ROI in which the brain activity is measured. ROIs can be defined either functionally (i.e. based on the pattern of activation) or anatomically (i.e. based on anatomic landmarks). Pure anatomical definitions often include areas that are not engaged by the task, and due to the normalization procedure activations of interest may lie outside the anatomically chosen ROI (Figure 5).

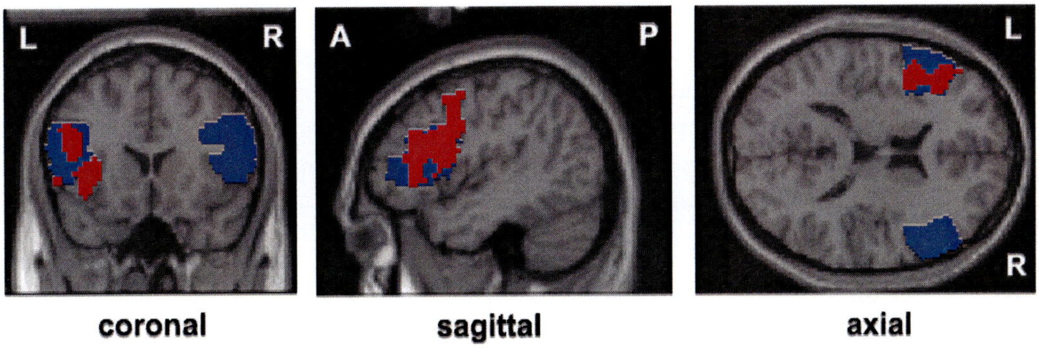

Figure 5. Here Broca's area and its homologue are defined anatomically as the opercular and triangular part of the inferior frontal gyrus (blue). Language associated brain activation within the dorsolateral prefrontal cortex of a selected subject is marked in red. This example shows that pure anatomic definitions of ROIs often include areas that are not engaged by the task, just as activations of interest might lie outside the ROI. Abbreviation used: L - left, R - right, A - anterior, P - posterior. (Figure modified from (Jansen et al. 2005)).

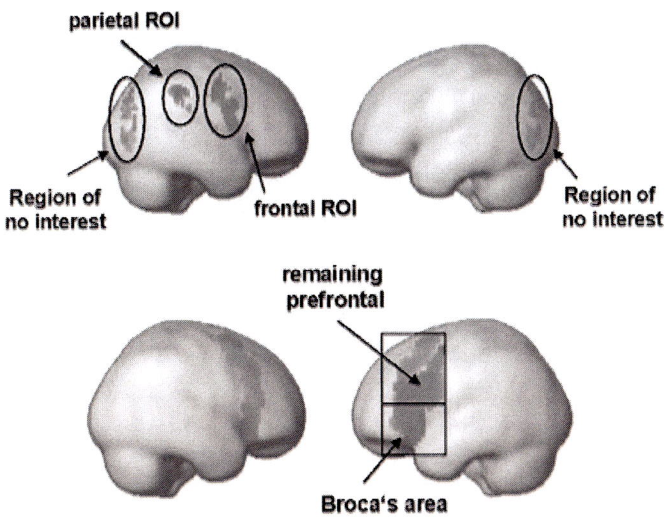

Figure 6. Top: Brain activation during the Landmark task. Regions of interest are defined functionally as the voxels in the superior parietal and the dorsolateral frontal cortex, respectively, activated in at least 70% of the subjects. Additional masking was needed to differentiate between regions of interest (frontal and parietal cortex) and regions of no interest (occipital cortex). Bottom: A differentiation between areas of interest and areas of no interest is not always as simple as in the previous example. During word generation, subjects activate a large region in the frontal cortex, encompassing Broca's area as well as other prefrontal regions. The borderline between both region remains to a certain degree arbitrary. (Figure modified from (Jansen et al. 2005)).

Furthermore, macroscopic anatomic landmarks are rarely reliable indicators of cytoarchitectonic borders (Amunts *et al.* 1999). Functional ROIs are easier to handle. They can be derived, for instance, from data of a pilot study investigating the functional activation in another group of subjects or in the same cohort with similar paradigms (Fernandez *et al.* 2001; Adcock *et al.* 2003). However, functionally defined ROIs can include regions outside the actual area of interest. Thus, they often require additional masking (Figure 6).

In summary, laterality calculation in functional neuroimaging depends on decisions concerning the effect size measure and a precise region-of-interest. Both can strongly bias the results. Their choices have to made clear, and it has to be shown which consequences they have on the results. Only when it has been shown that a lateralization index delivers reproducible results in the normal population, can it be used as a useful marker to investigate abnormalities in the diseased brain.

Acknowledgments

This work was supported by the NRW-Nachwuchsgruppe Kn2000 of the Nordrhein-Westfalen Ministry of Education and Research (Fö.1KS9604/0), the Interdisciplinary Center of Clinical Research Münster (IZKF Projects FG2 and Kne3/074/04), the Innovative Medizinische Forschung Münster (KN520301), the Deutsche Forschungsgemeinschaft (Kn 285/6-1 and 6-3) as well as a EC Marie Curie Research and Training grant (RTN:LAB 512141).

References

Adcock JE, Wise RG, Oxbury JM, Oxbury SM, Matthews PM. 2003. Quantitative fMRI assessment of the differences in lateralization of language-related brain activation in patients with temporal lobe epilepsy. *Neuroimage* 18: 423-438.

Amunts K, Schleicher A, Burgel U, Mohlberg H, Uylings HB, Zilles K. 1999. Broca's region revisited: cytoarchitecture and intersubject variability. *J. Comp. Neurol.* 412: 319-341.

Annett M, Alexander MP. 1996. Atypical cerebral dominance: predictions and tests of the right shift theory. *Neuropsychologia* 34: 1215-1227.

Bell BD, Davies KG, Haltiner AM, Walters GL. 2000. Intracarotid amobarbital procedure and prediction of postoperative memory in patients with left temporal lobe epilepsy and hippocampal sclerosis. *Epilepsia* 41: 992-997.

Bertolino A, Blasi G, Caforio G *et al.* 2004. Functional lateralization of the sensorimotor cortex in patients with schizophrenia: Effects of treatment with olanzapine. *Biological Psychiatry* 56: 190-197.

Binder JR, Swanson SJ, Hammeke TA *et al.* 1996. Determination of language dominance using functional MRI: a comparison with the Wada test. *Neurology* 46: 978-984.

Broca P. 1861a. Perte de la Parole, Ramollissement Chronique et Destruction Partielle du Lobe Antérieur Gauche du Cerveau. *Bulletin de la Société Anthropologique 2*: 235-238.

Broca P. 1861b. Remarques sur le siége de la faculté du langage articulé, suivies dúne observation d'aphemie. *Bulletin de la Société Anatomique* 6: 330-357.

Cohen MS, DuBois RM. 1999. Stability, repeatability, and the expression of signal magnitude in functional magnetic resonance imaging. *J. Magn. Reson. Imaging.* 10: 33-40.

Corballis MC. 1992. On the evolution of language and generativity. Cognition 44: 226.

Deppe M, Ringelstein EB, Knecht S. 2004. The investigation of functional brain lateralization by transcranial Doppler sonography. *Neuroimage* 21: 1124-1146.

Fernandez G, de Greiff A, von Oertzen J *et al.* 2001. Language mapping in less than 15 minutes: real-time functional MRI during routine clinical investigation. *Neuroimage* 14: 585-594.

Fernandez G, Weis S, Stoffel-Wagner B *et al.* 2003. Menstrual cycle-dependent neural plasticity in the adult human brain is hormone, task, and region specific. *J. Neurosci.* 23: 3790-3795.

Floel A, Ellger T, Breitenstein C, Knecht S. 2003. Language perception activates the hand motor cortex: implications for motor theories of speech perception. *Eur. J. Neurosci.* 18: 704-708.

Friederici AD, Alter K. 2004. Lateralization of auditory language functions: A dynamic dual pathway model. *Brain and Language* 89: 267-276.

Friston KJ, Harrison L, Penny W. 2003. Dynamic causal modelling. *Neuroimage* 19: 1273-1302.

Frost JA, Binder JR, Springer JA *et al.* 1999. Language processing is strongly left lateralized in both sexes. Evidence from functional MRI. *Brain* 122: 199-208.

Garcia RR, Montiel JF, Villalon AU, Gatica MA, Aboitiz F. 2004. AChE-rich magnopyramidal neurons have a left-right size asymmetry in Broca's area. *Brain Research* 1026: 313-316.

Heeger DJ, Ress D. 2002. What does fMRI tell us about neuronal activity? *Nature Reviews Neuroscience* 3: 142-151.

Heim S, Keil A. 2004. Large-scale neural correlates of developmental dyslexia. *Eur. Child Adolesc. Psychiatry* 13: 125-140.

Hier DB, LeMay M, Rosenberger PB. 1979. Autism and Unfavorable Left-Right Asymmetries of the Brain. *Journal of Autism and Developmental Disorders* 9: 153-159.

Hirata M, Kato A, Taniguchi M *et al.* 2004. Determination of language dominance with synthetic aperture magnetometry: comparison with the Wada test. *Neuroimage* 23: 46-53.

Jansen A, Floel A, Deppe M *et al.* 2004. Determining the Hemispheric Dominance of Spatial Attention - A Comparison Between fTCD and fMRI. *Hum. Brain Mapp.* 23: 168-180.

Jansen A, Menke R, Förster AF *et al.* 2005. *Reproducibility, information content and robustness of quantitative fMRI measures for the lateralization of cognitive brain functions.* (in preparation).

Kennan RP, Kim D, Maki A, Koizumi H, Constable RT. 2002. Non-invasive assessment of language lateralization by Transcranial near infrared optical topography and functional MRI. *Human Brain Mapping* 16: 183-189.

Knake S, Haag A, Hamer HM *et al.* 2003. Language lateralization in patients with temporal lobe epilepsy: a comparison of functional transcranial Doppler sonography and the Wada test. *Neuroimage.* 19: 1228-1232.

Knecht S, Deppe M, Ebner A *et al.* 1998. Noninvasive determination of language lateralization by functional transcranial Doppler sonography: a comparison with the Wada test. *Stroke* 29: 82-86.

Knecht S, Drager B, Deppe M *et al.* 2000. Handedness and hemispheric language dominance in healthy humans. *Brain* 123 Pt 12: 2512-2518.

Knecht S, Drager B, Floel A *et al.* 2001. Behavioural relevance of atypical language lateralization in healthy subjects. *Brain* 124: 1657-1665.

Knecht S, Floel A, Dräger B *et al.* 2002. Degree of language lateralization determines susceptibility to unilateral brain lesions. *Nat. Neurosci.* 5: 695-699.

Knecht S, Jansen A, Frank A *et al.* 2003. How atypical is atypical language dominance? *Neuroimage.* 18: 917-927.

Kosslyn SM. 1999. *If neuroimaging is the answer, what is the question?* Philosophical Transactions of the Royal Society of London Series B-Biological Sciences 354: 1283-1294.

Lutkenhoner B. 2003. Magnetoencephalography and its Achilles' heel. *J. Physiol. Paris* 97: 641-658.

McManus IC. 1985. Handedness, language dominance and aphasia: a genetic model. *Psychol. Med. Monogr. Suppl.* 8: 1-40.

Mostofsky SH, Cooper KL, Kates WR, Denckla MB, Kaufmann WE. 2002. Smaller prefrontal and premotor volumes in boys with attention-deficit/hyperactivity disorder. *Biological Psychiatry* 52: 785-794.

Nagata SI, Uchimura K, Hirakawa W, Kuratsu JI. 2001. Method for quantitatively evaluating the lateralization of linguistic function using functional MR imaging. *AJNR Am. J. Neuroradiol.* 22: 985-991.

Pease A, Pease B. *Why Men Don't Listen and Women Can't Read Maps.* 1998.

Pujol J, Deus J, Losilla JM, Capdevila A. 1999. Cerebral lateralization of language in normal left-handed people studied by functional MRI. *Neurology* 52: 1038-1043.

Ramsey NF, Sommer IE, Rutten GJ, Kahn RS. 2001. Combined analysis of language tasks in fMRI improves assessment of hemispheric dominance for language functions in individual subjects. *Neuroimage* 13: 719-733.

Rizzolatti G, Arbib MA. 1998. Language within our grasp. *Trends Neurosci.* 21: 188-194.

Rossion B, Joyce CA, Cottrell GW, Tarr MJ. 2003. Early lateralization and orientation tuning for face, word, and object processing in the visual cortex. *Neuroimage* 20: 1609-1624.

Sabsevitz DS, Swanson SJ, Hammeke TA *et al.* 2003. Use of preoperative functional neuroimaging to predict language deficits from epilepsy surgery. *Neurology* 60: 1788-1792.

Specht K, Willmes K, Shah NJ, Jancke L. 2003. Assessment of reliability in functional imaging studies. *Journal of Magnetic Resonance Imaging* 17: 463-471.

Stephan KE. 2004. On the role of general system theory for functional neuroimaging. *Journal of Anatomy* 205: 443-470.

Vogel JJ, Bowers CA, Vogel DS. 2003. Cerebral lateralization of spatial abilities: A meta-analysis. *Brain and Cognition* 52: 197-204.

Votaw JR. 1995. The Aapm/Rsna Physics Tutorial for Residents. *Radiographics* 15: 1179-1190.

Vuust P, Pallesen KJ, Bailey C *et al.* 2005. To musicians, the message is in the meter Pre-attentive neuronal responses to incongruent rhythm are left-lateralized in musicians. *Neuroimage* 24: 560-564.

Wada J, Rasmussen T. 1960. Intracarotid injection of sodium amytal for the lateralization of cerebral speech dominance. *J. Neurosurg.* 17: 266-282.

Weiss EM, Hofer A, Golaszewski S *et al.* 2004. Brain activation patterns during a verbal fluency test - a functional MRI study in healthy volunteers and patients with schizophrenia. *Schizophr. Res.* 70: 287-291.

Wildgruber D, Riecker A, Hertrich I *et al.* 2005. Identification of emotional intonation evaluated by fMRI. *Neuroimage* 24: 1233-1241.

In: Focus on Brain Research
Editor: Casey J. Resch, pp. 123-141
ISBN 978-1-60021-395-3

Chapter V

EEG-MAPPING STUDY OF SEX DIFFERENCES DURING VISUAL CREATIVE THINKING

Nina V. Volf, Olga M. Razumnikova and Irina V. Tarasova
State Research Institute of Physiology of Siberian Department of
Russian Academy of Medical Sciences,
Novosibirsk, Russia

ABSTRACT

The study investigated sex-related differences in cortical activity during performance of a figural creative task: generating visual images on the base of a presented fragment. Additionally, we were interested in whether there are sex differences in the influence of instructions "to create images" (i) or "to create unique images" (ui) on the creativity scores and EEG-patterns. The study is based on 26 students (13 males and 13 females). EEG was recorded from 16 sites. The spectral power density and EEG coherence were computed for six frequency ranges between 4-30 Hz. Under (ui) vs (i) instruction fluency score decreased in both groups while originality scores insignificantly increased. Sex differences between power changes during imagery task were found only in the theta1,2-bands. Independent of instruction, power decreased more in the right than in the left hemisphere in the female group; no hemispheric differences were found in males. Sex differences in task induced changes of coherence were most distinct in the theta and beta bands. In the theta bands, these changes manifest themselves mainly as increases of coherence in females but decreases in males. Under (i) instruction figural thinking was characterized by increased coherence over the left hemisphere and over the left frontal regions paired with right posterior regions in females in comparison to males. Under (ui) instruction sex differences were mainly presented in indices of interhemispheric cooperation. Changes in the beta frequency bands manifest themselves as more coherence increase in males in comparison with females. In the beta1, under (i) instruction males vs. females demonstrated increased coherence in the left frontal and between left and right frontal regions. The interhemispheric coupling between posterior areas in the beta1 seems to be stronger in males than in females especially under (ui) instruction. In the beta2 frequency band only in male group, diffuse intra- and interhemispheric short- and long–distance coherence increases were found under both instructions. This effect was more pronounced under (i) instruction, when task solutions were not limited by originality requirements. Our results indicate that brain mechanisms engaged by men and women during creative figural thinking are different.

INTRODUCTION

In order to differentiate the creative style of thinking Y.P. Guilford used the descriptive term 'divergent' thinking (i.e. original, multivalent) in contrast to 'convergent' one (i.e. unambiguous, stereotyped). There are theoretical reasons for separating mental abilities into separate components: verbal and visual- spatial processes (Logie, 1995; Shah and Miyake, 1996; Halpern, 2000). This had led to the development of investigations designed for study of verbal and figural creativity.

It is well established now that verbal and visual- spatial cognitive abilities are the loci of sex differences. (Maccoby and Jacklin,1974; Linn and Petersen, 1985; Hyde and Linn, 1988; Voyer et al., 1995; Halpern, 2000). Psychological studies have found better visual-spatial abilities in men than women. However, it should be mentioned that analysis of different visual-spatial tests (Linn and Petersen, 1986; Halpern, 2000) have broken spatial tasks into different types. It is generally agreed that the largest and most consistent findings of sex differences in spatial ability occur in studies of mental rotation (Linn and Petersen 1986, Voyer et al., 1995). Men as a group consistently outperform women on tests assessing this skill. However, in agreement with the point of view that visual ability is not a homogenous construct, gender differences may strongly depend on which component of visual spatial ability is measured.

Figural tasks of Torrance Tests of Creative Thinking (TTCT) are divergent visual tasks which are widely used in research of creativity and imaginative skills. In contrast to mental rotation tasks, many studies reported higher scores for female in comparison to male students on figural creativity (Warren and Luria, 1972; Shutiva, 1991; Stephens et al., 2001). In this regard, Farah (1989) showed that different components of imagery processing such as mental imagery and mental rotation are differently lateralized with the generation of mental images from memory depending primarily upon structures in the posterior left hemisphere, and the rotation of mental images depending primarily upon structures in the posterior right hemisphere. Recent studies also have found neurophysiological evidences for greater hemispheric interconnections and a less pronounced brain asymmetry during creative mental processes (Petsche and Etlinger, 1998; Athley et al., 1999; Carlsson et al., 2000; Razoumnikova, 2000). At the same time, it has been known that women have a less pronounced hemispheric lateralization along with more flexible thinking strategies in comparison with men (Davidson, et al., 1976; Kimura, 1987; Hausmann and Gunturkun, 1999).

A large body of morphological, neurochemical and neurophysiological data of the sex dimorphism in hemispheric organization (Kimura, 1987; Shaywitz et al., 1995; Zaidel et al., 1995; Volf and Razumnikova, 1999) causes us to anticipate an occurrence of gender differences in hemispheric organization of visual-spatial functions. Indeed, a number of electrophysiological studies reported sex differences in task-related EEG power and coherence changes during a mental rotation task (Recher and Rappelsberger,1999; Roberts and Bell, 2000; 2002). Using similar EEG technology sex related differences in the EEG-patterns were found during successful performance of a verbal creative task (Razumnikova, 2004). Creative men were characterized by massive increases of amplitude and interhemispheric coherence in the beta2 whereas creative women showed more local increases of the beta2-power and coherence. On the contrary, the task-induced desynchronization of the alpha1 rhythm in creative women was topographically more expanded as compared with men who demonstrated greater interhemispheric coherence than women did. Nevertheless, the problem of gender-related differences in brain organization during creative thinking is far from resolved. Thus, the question whether different cortical networks are engaged in men and women during figural creative thinking is still unexplored.

In this respect, the present study was designed to investigate gender-related differences in behavioral measures and EEG power and coherence during performance of figural divergent tasks. As a number of studies reported the influence of instruction to be original on participants' divergent thinking scores (Harrington, 1975; Owen, Baum; 1985; Chand, Runco, 1993), we studied a test induced EEG reactivity under the different instructions.

METHOD

Subjects

26 (13 females and 13 males) right-handed students from Novosibirsk State University and from Novosibirsk Technical University participated in the study. Ages ranged from 19 to 23 years. Handedness was verified by a handedness questionnaire (Annett, 1970). All the subjects signed informed consents to participate in the study, and the research received approval of the institutional ethics-committee.

Materials and Procedure

Ten figural tasks from the Torrance Tests of Creative Thinking were used (Torrance, 1984). A figural task was to create as many pictures as possible based on an incomplete figure. The experiment consisted of two parts. These parts differed in instructions given to the subjects. In the first part, the instruction was "to create images"(i), in the second part it was "to create unique images" (ui). Five tasks with different incomplete figures were presented under (i) instruction, and next the five tasks under (ui) instruction.

Subjects were sitting in a comfortable chair with head supported in a semi-darkened, sound - attenuated room. Each experimental part began with the 1-min presentation of a fixation point in the center of the video monitor situated in front of the subject. After that instead of a fixation point an incomplete figure was displayed for a period of 1 min. In this period, participants had to mentally create pictures using the figure presented on the screen.

The verbal response (the name of the picture) had to be given for every completed image. Each task was followed by a 1 min pause in which a fixation point was presented on the computer screen. In an attempt to avoid artifacts, subjects were asked to maintain fixation by looking on the figure during the task performance and by looking on the fixation point during the intertask interval. After the end of the EEG experiment, the electrode cap was taken off and students had to draw on a blank all images created during the experiment (Figure 1).

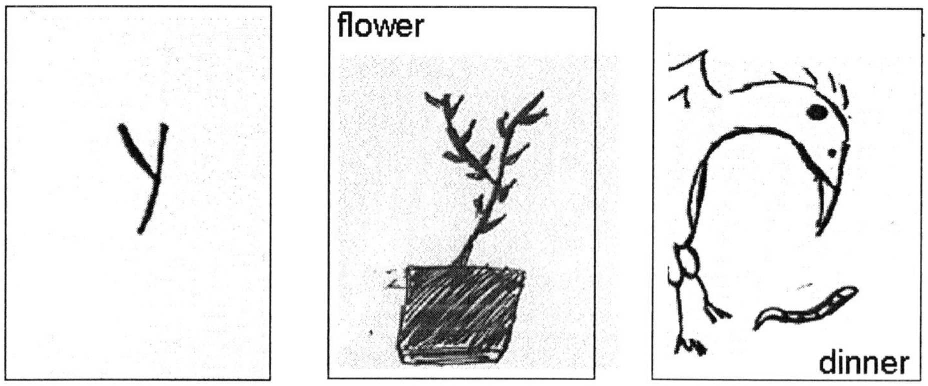

Figure 1. Examples of incomplete figure (left) and created stereotype (middle, originality score-0,09) and original (right, originality score -1) pictures.

Performance in each task was scored with the help of an original computer program. Originality scores were obtained on the basis of the normative sample that was composed of 342 male and female students aged 19-23 years tested during two years prior to the experiment. Each response could be attributed a score 1/(N+1) where N is the occurrences of the idea in the sample. So, high scores correspond to rare ideas (the highest score 1 corresponds to the unique idea). The number of ideas represents fluency. The creativity index was then computed separately for (i) and (ui) conditions by adding the originality scores for all ideas.

EEG Recordings and Analysis

EEG was recorded from 16 scalp sites according to the international 10/20 system: Fp1, Fp2, F3, F4, F5, F6, T3, T4, T5, T6, C3, C4, P3, P4, O1, O2. Electrodes were attached by stretchable electrode cap and referenced to linked earlobes, and a ground electrode was attached to the centre of the forehead. The EEG was amplified and filtered by Galileo electroencephalograph with a bandpass filter of 0.3 - 30 Hz. Electrode resistance was maintained below 5 kΩ. EEG data were digitized at a sampling rate of 256 Hz and stored on a hard disk. Prior to analysis, the individual EEG traces were manually inspected to be free of eye-movement artifacts and to contain no electromyographic activity. One EEG epoch consisted of 2 sec. Epochs were overlapped by 50 %. 29 artifact-free epochs per subject were obtained for every task record and for every resting period. Artifact-free epochs of EEG containing the figure presentations were Fourier transformed, using a Hanning window, to compute averaged power and cross-power spectra. Artifact-free epochs of resting EEG were also processed and the corresponding spectral parameters served as 'baseline' for the task condition. Coherence was defined as $C(f)^2=(S_{xy}(f))^2/(S_{xx}(f)S_{yy}(f))$ where $S_{xy}(f)$ is the cross-power spectral density and $S_{xx}(f)$, $S_{yy}(f)$ are respective autopower spectral densities. The averaged power density (μV^2 /Hz) and coherence for each of the six frequency bands: theta1(4-6 Hz), theta2 (6-8 Hz), alpha1 (8-10 Hz), alpha2 (10-13 Hz), beta1 (13-20 Hz), and beta2 (20-30 Hz) were calculated. Fourier transformation and power and coherence calculations were done using "Neuromapping 3.55 (MBN, Russia)".

The selection of lower and upper theta, alpha and beta bands seemed most promising because it is well established that the frequency components of all these bands are sensitive to task demands (Petsche, 1996; Petsche, et al., 1997; Klimesch et al., 1997; Volf and Razumnikova, 1999; Razumnikova, 2004). The delta band was excluded from the analysis on account artifacts in many subjects.

Statistical Analysis

Statistical analysis of EEG data was done using the package Statistica for Windows, StatSoft (1995). Power data from each electrode were log-transformed and coherence values were Fisher-z-transformed in order to normalize the distribution. For (i) and (ui) instructions and for each frequency band separately mean baseline and mean experimental power and coherence values were obtained by averaging values obtained during 5 resting periods and during 5 test conditions for each electrode in each subject. Task-related power and coherence reactivity were calculated by subtraction of the baseline values from values obtained during figural task performance. Grand mean values were obtained by averaging coherence reactivity scores across electrode pairs (for intrahemispheric and interhemispheric coherence separately).

We applied repeated measure ANOVAs to task-related power reactivity scores and to the grand mean coherence reactivity values in order to test main factors and their interactions. The Greenhouse-Geisser procedure was adopted in order to compensate for non-sphericity. For the evaluation of all possible single coherence differences between the task and the resting EEG, paired T-tests were calculated separately for each of the six frequency bands in (i) and (ui) conditions for male and female subjects. T–test for independent samples were applied to evaluate between group differences.

Considering amount of data (120 variables for band) being subjected to T-testing, the risk of a false alarm or type I error is high. However, statistical correction of the P values for such a large number of variables may result in failed alarm or type II errors. So, for statistical reasons these evaluations are only of describing nature and focus is directed on electrodes involved in numerous coherence changes ("centers of gravity" or "processing nodes") opposite to single change (Abt, 1988; Lacroix, et al., 1995; Petsche, 1996; Aftanas et al., 1998).

RESULTS

Behavioral Data

Repeated measures ANOVAs were performed on creativity, originality and fluency scores in order to exanimate effects of the factors GENDER (males vs. females), and INSTRUCTION ((i) vs. (ui)). The main effect of factor INSTRUCTION was found for fluency scores ((F(1,24)=8,66; P=0,007). Fluency score was higher for the (i) than for (ui) instruction. Furthermore, there was a statistical tendency (P=0,07) for a higher originality score under (ui) than (i) instruction (Figure 2). For creativity scores there were no main effects and no significant interactions.

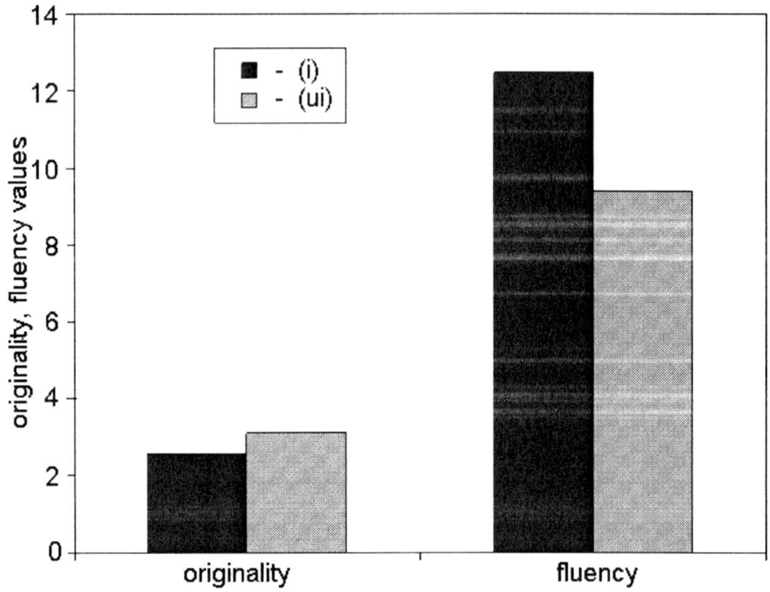

Figure 2. Indices of originality and fluency under instructions "to create images" (i) or "to create unique images"(ui). The same signs indicate significant differences for respective values.

Power

In order to determine differences in task-related power reactivity between two gender groups ANOVA was calculated for repeated measures - GENDER x INSTRUCTION x ELECTRODE LOCATION (8 positions) x HEMISPHERE (left vs. right). Gender differences were found only in theta 1,2 frequency bands. Two-way interaction between GENDER and HEMISPHERE was found in theta1 (F(1,24)= 6,30; P=0,019) and theta2 (F(1,24)=4,77; P=0,039) bands. Inspection of the respective means of the interactions revealed that females demonstrated greater desynchronization in the right hemisphere than in the left one. In males no significant difference between right and left hemisphere was found (Figure 3).

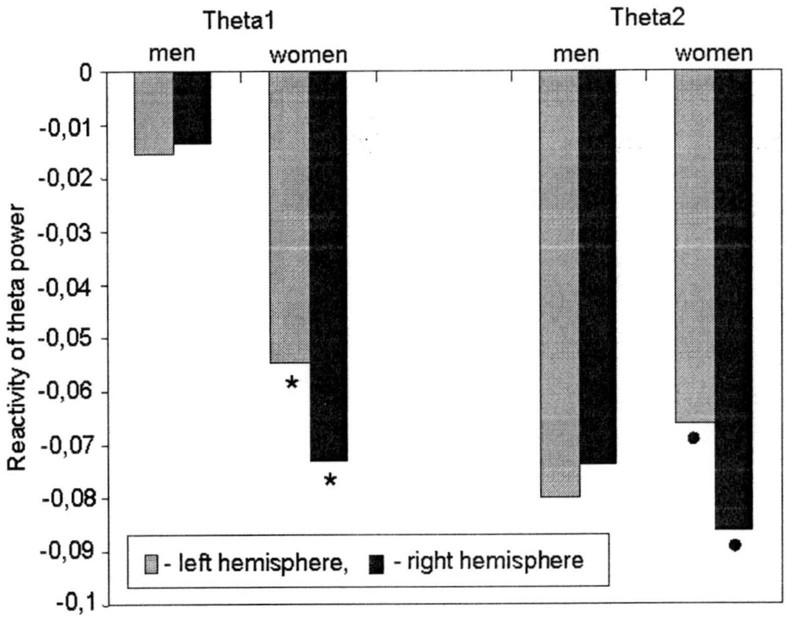

Figure 3. Sex differences in the hemispheric asymmetry of power reactivity (log theta1 and theta2 power relative to baseline) during visual creative thinking. The same signs indicate significant results of planed comparisons for respective values.

Coherence

Grand mean coherence. Repeated measures ANOVAs were performed on mean coherence reactivity values in order to examine effects of the factors GENDER, FREQUENCY BAND (six bands), COHERENCE (intra- vs. interhemispheric) and INSTRUCTION. Main FREQUENCY BAND effect was significant (F(5,120)=8,60; P=0,000). Inspection of respective means revealed task-induced coherence increases in beta1,2 bands whereas mean coherence values were diminished across other frequency bands. The interaction between GENDER and FREQUENCY BAND (F(5,120)=6,43; P=0,000) yielded significant frequency-dependent coherence reactivity differences between gender groups. The respective means of the interaction revealed that men demonstrated higher than women coherence reactivity in beta1 and beta2 bands while in theta1 and theta2 bands coherence reactivity in women was higher than in men. (Figure 4). The interaction between COHERENCE and FREQUENCY BAND (F(5,120)=2,85; P=0,018) indicated that differences between intra- and interhemispheric mean coherence reactivity are frequency specific. Only in theta1 frequency band, intrahemispheric reactivity value was higher than interhemispheric reactivity value. For other frequency bands, no significant differences between intra- and interhemispheric mean coherence reactivity were found. INSTRUCTION x FREQUENCY BAND (F(5,120)=2,55; P=0,031) interaction was significant. Between-instruction differences were pronounced for alpha2 and beta1 bands due to higher coherence reactivity under (i) in comparison with (ui) instruction.

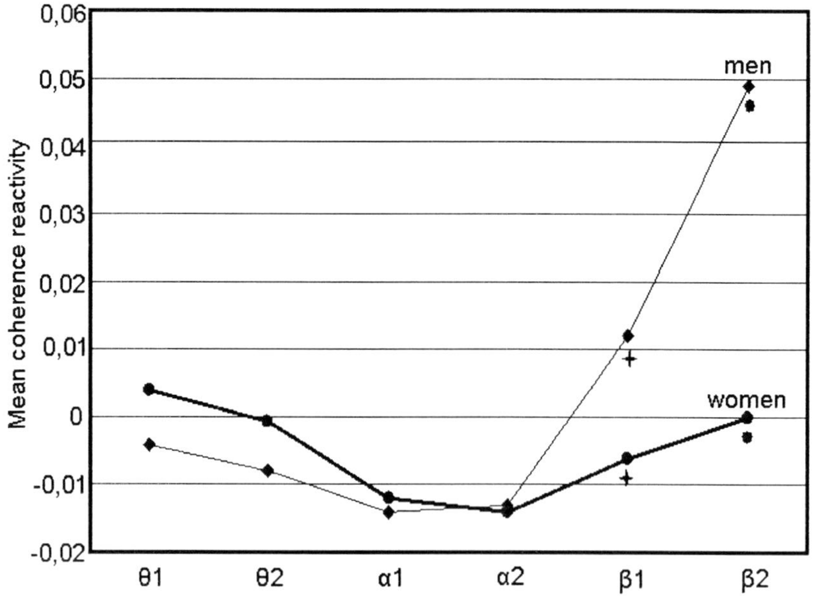

Figure 4. Differences of mean coherence reactivity (average of all electrode pairs) between men and women. The same signs indicate significant results of planed comparisons for respective values.

Topographic Results

To get an idea of the overall topographic pattern of the main coherence differences elicited by the figural task in male and female subjects, comparisons of EEG coherence during task performance with the parameters of EEG at rest were made for both (i) and (ui) instructions. In order to determine between-gender differences T-test for independent samples was applied to coherence reactivity scores. The resulting error probabilities ($p<0,05$) were mapped onto brain schemata as connecting lines between the corresponding electrodes and had to be considered as descriptive only.

In consequence with analysis of grant means of coherence, clear topographic coherence differences between males and females were found in theta and beta frequency bands. In alpha bands gender-related differences were found for less than 5% of electrode pairs for each instruction. Therefore, in order to illustrate patterns of coherence changes for males and females compared with EEG at rest, only brain maps of theta1,2, and beta 1,2, bands are shown both for the (i) and the (ui) conditions.

As the theta1 band is concerned, the figural task produced predominantly coherence decreases in males and mainly coherence increases in females. Decreases of coherence were more noticeable in (i) condition. As can be seen in Figure 5, decupling was mainly related to long-distance intra- and interhemispheric connections between occipital and frontal brain areas in males and to short-range connections between fronto-temporal regions of the right hemisphere in females. Probability mapping of local coherence differences between male and female subjects showed the left hemisphere frontal nodes (Fp1, F7) for many changes with posterior sites of both hemispheres. Females in comparison with males showed greater cooperation between abovementioned areas.

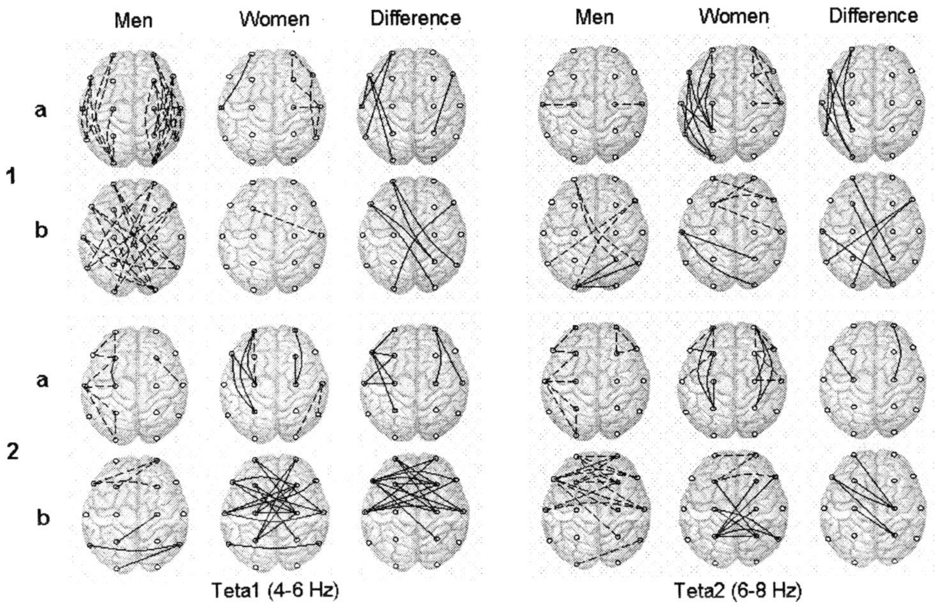

Figure 5. Coherence changes (intrahemispheric – a and interhemisperic –b) in men and women while a pictures were mentally creating compared to the baseline EEG. 1) Instruction "to create images"; 2) instruction "to create unique images". Full lines mean coherence increase, broken lines mean coherence decrease compared to baseline EEG. The scheme on the right side (difference) demonstrate coherence reactivity differences between men and women. Full lines mean higher task induced coherence reactivity in women as compared with men.

Under (ui) instruction large number of coherence increases in comparison with baseline data was found in female subjects. Intrahemispheric coherence increase was distinct in the left hemisphere of females, whereas in males task was accompanied by coherence decrease in the same hemisphere. Males presented also a few interhemispheric coherence decreases in which the frontal electrodes participate and few coherence increases between posterior electrode sites of opposite hemispheres. In females compared with the baseline condition, the figural task provoked a large number of interhemispheric coherence increases mainly between frontal and centro-temporal regions. As coherence differences map for (ui) instruction shows, the general trend was that female subjects in comparison with males displayed enhanced short distance cooperation between frontal brain areas mainly in the left hemisphere (Fp1, F3, F7, T3, C3) and greater interhemispheric cooperation between frontal central and temporal electrode sites (Fp1, Fp2, F3, F4, F5, F6, T3, T4, C3,C4).

In the theta2 band, both male and female subjects showed as coherence increases, so and coherence decreases. In so doing, males showed more coherence decreases while females showed more coherence increases. Under (i) instruction, compared with baseline condition males showed a few intra- and interhemispheric coherence decreases along with increase of interhemispheric coupling with a 'processing node' in the left occipital region (O1). Females showed coherence increases in the left hemisphere with a 'processing node' in the parietal site (P3) and enhancement of interhemispheric coherence between a few posterior electrode sites, whereas right frontal inter-regional coherence decreased in parallel with frontal interhemispheric decoupling. In comparison with males, females demonstrated greater cooperation between fronto-temporal, fronto-parietal and fronto-occipital electrodes in the left hemisphere. Females also displayed higher interhemispheric cooperation between frontal and parieto-occipital areas. In (ui) condition, in comparison with baseline, males showed intra- and interhemispheric functional decoupling. Intrahemispheric decoupling was more extensive between the areas in the left hemisphere. In the right one, decrease of cooperation was pronounced only between frontal electrode sites. Interhemisheric coherence decreases were diffuse and involved mainly frontal and anterior temporal regions. In contrast to males, females demonstrated increases of coherence between parietal and frontal sites in the left and in the right hemispheres with respect to baseline EEG. In the same time, the short-range coherence between frontal electrode sites decreased in female subjects. They also showed closer interhemispheric cooperation than at rest with the "center of gravity" of coherence increases at P3 site. As revealed in Figure 5 (right column), females seem to need closer cooperation between the right centro-parietal and the left frontal parts of the brain than males.

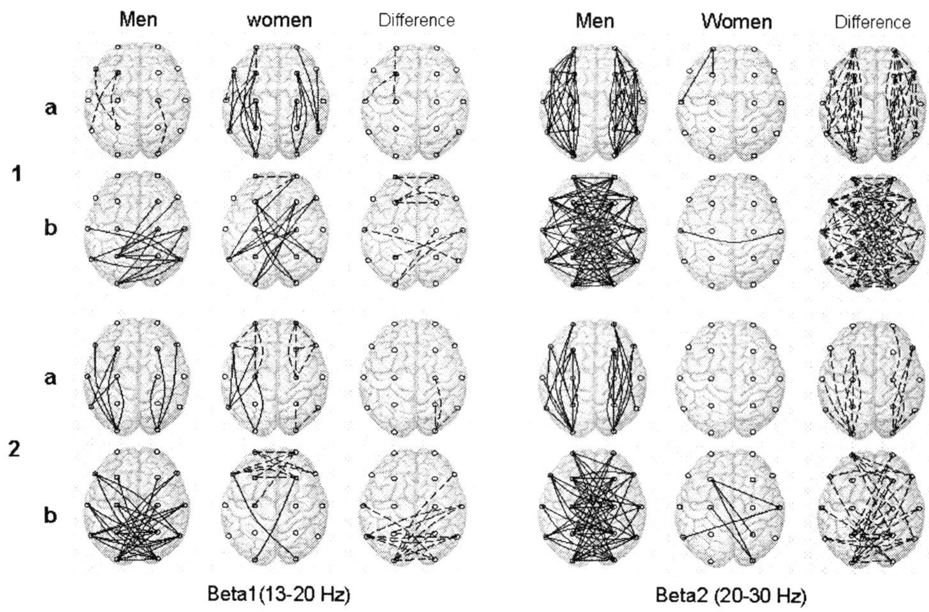

Figure 6. Coherence changes in men and women while a pictures were mentally creating with respect to the EEG at rest in the beta bands. For "difference" broken lines mean higher task induced coherence reactivity in men as compared with women. For further description see legend of Figure 5.

In beta1, under (i) instruction, mainly task-induced coherence increase was found in both groups with, however, different topographical representations. For intra- and interhemispheric coherence increases (Figure 6, left and middle column) the main "center of gravity" emerged at left parietal site (P3) in males and at left posterior temporal site (T5) in females. As for gender-related differences, the coherence changes between left fronto-centro-temporal electrode sites are the most ones mentioned. Concerning interhemispheric coherence changes, males in comparison with females demonstrated closer functional coupling between frontal parts of different hemispheres as well as between their posterior parts.

In (ui) condition, in males interhemispheric coherence increases in both hemispheres concerned mainly O1 and O2 electrode sites. The similar coherence increases were found in the left hemisphere of females. In contrast to this, only in females coherence decreased between frontal regions in both left and right hemispheres and between electrode sites in the right posterior region. For interhemispheric coherence, males showed diffuse coherence increases between numerous positions with the exception of Fp1 and Fp2 electrode sites. In females, coherence decrease was observed between frontal regions of different hemispheres. As the right column demonstrates there are gender dependent differences in changes of coherence between posterior regions contralaterally and homolaterally in the right hemisphere.

In the case of beta2 frequency band (Figure 6, left column) under both (i) and (ui) instructions in males, visual tasks elicited diffuse intra- and interhemispheric coupling, involving most regions. These effects were more pronounced under (i) instruction. In contrast, female subjects did not show any intrahemispheric coherence increases under both instructions. As for significant interhemispheric changes, most worth mentioning are the coherence increases under (ui) instruction with the processing node in F3. Under (i) instruction, gender-related differences were presented as higher intra- and interhemispheric cooperation between numerous electrode sites in males. Under (ui) instruction males vs. females were characterized by the presence of the left and right occipital focuses of increased intrahemispheric coherences and by closer interhemispheric cooperation involving diffuse regions of the entire right hemisphere whereas in the left hemisphere interhemispheric coherence increases concentrated in the Fp1, P3 and O1 electrode sites.

CONCLUSION

Analysis of the behavioral measures revealed no sex differences in the efficiency of performance of figural creative tasks in our study. These results vary from a number of studies that showed higher scores in females in comparison to males (Warren and Luria, 1972; Shutiva, 1991; Stephens et al., 2001). In this connection, it should be noted that female advantage on creative tasks was found in preadolescent and adolescent children, while university students participated in our study. So, age differences may be responsible for reported discrepancies. Another source of the differences may be academic ability. It was found that there were no gender differences on the figural component of the TTCT in high-achieving adolescents (DeMoss et al., 1993). Our data are in consequence with this result as in comparison with unselected school-age participants, university students may be considered as a high academic ability group.

The results of the present study exhibited EEG power and coherence differences between male and female subjects during performance of a figural creative task.

The analysis of power measures indicated that females demonstrated greater theta desynchronization in the right hemisphere than in the left one. In males no significant difference between ther right and the left hemisphere theta desynchronization was found. The present results contrast with the increase of theta power reported in relation with task difficulty (Petche et al., 1986; Lang et al., 1988; Gundel and Wilson, 1992; Inouye et al., 1993), encoding and retrieval in memory tasks (Klimesh et al., 1997), emotional factors (Simonov, 1987) and attention (Laukka 1995, Stroganova, 1998). In comparison, the decrease of theta power was found when subjects had to remember the spatial features of a cue stimulus along the delay period (Bastiaansen et al., 2002; Babiloni et al., 2004). Furthermore, decrease of the theta oscillations were reported during presentation of affective negative information (Aftanas, 2000). Authors of these articles supposed that theta desyncronization was possibly related to inhibition of external attention and concentration on internal processing. Thus, one may assume that during visual imagery the decreased theta power is associated with concentration on mental processes and maintenance of visual images.

In the present study, sex differences in theta asymmetry were found. This is in line with the study of Gur and collaborators (Gur et al., 1982) who in an rCBF study found stronger laterality effect for a spatial task among right-handed females. However, the present results are in disagreement with the absence of sex differences in rCBF asymmetries reported by other authors during performance of spatial tasks (Deutsch et al., 1988; Wendt and Risberg, 1994). It has been suggested that "spatial ability" is not one but several different abilities (Linn and Petersen, 1985; Halpern, 2000). It is probable that the aptness to form subjective visual representations, does not coincide with the ability to spatially manipulate objects. For instance, Carlsson and collaborators (2000) found that a highly creative group was not better on the spatial tests. So, sex differences related to visuo-spatial abilities might be a more complex issue than previously considered.

In line with recent suggestions, coherence changes may be considered as an indicator of the inter-relationship (information flow) between brain regions (Thatcher et al., 1986; Braitenberg and Schuz, 1991). Increases of coherence suggest a close cooperation of the two regions in question in the respective frequency band. Decreases of coherence show that two regions became more separated or that increased processing between a given area and subcortical areas takes place. In any case, the number of coherence changes centered around an electrode has proved to be an indicator of the functional importance of this region ("center of gravity" or "processing node") for the task (Petsch, 1996; Weiss and Rappelsberger, 1996; Petsch et al., 1997).

This study shows that sex differences in task-induced changes of electrical relationships between different cortical regions are most distinct in the theta and beta frequency bands.

In the theta bands these changes manifest themselves mainly as significant increases of coherence in females but as significant coherence decreases in males. It is worth mentioning that in all participants under (i) instruction intrahemispheric coherence decreases are more numerous in the right in comparison with the left hemisphere. In contrast, the left sided coherence increases were found only in female group. As result, females displayed more cooperation between areas of the left hemisphere than did males. It could be supposed that gender-related differences in the left hemisphere involvement during the task is related to the way in which females and males solve the problem. Kosslyn and collaborators (1992; 1995 a, b) provide evidence that visual mental images can be formed in two ways. One method relies on stored descriptions to arrange parts of image objects, and the other relies on stored information about metric distances to arrange parts. They assumed that spatial relations are stored using languagelike, symbolic representations being processed more effectively in the left hemisphere, and the other being processed better in the right hemisphere. Studies of the neural substrate of mental image generation with PET and functional MRI have revealed considerable evidence for the greater activity in the right than in the left hemisphere during different tasks including basic visuo-spatial analysis (McIntosh et al., 1994; Wend and Risberg, 1994) although left hemisphere was more involved in generation of visual images of the words' referents (D'Esposito et al., 1997). It was also shown that construction of visual images from geometric figures is associated with right hemisphere functional integration in participants with artistic thinking and with the left sided integration in participants with abstract thinking (Ivanitzkii et al., 1990). Thus, one may assume that women solved figural creative problem more by relying on the symbolic representations. Females in comparison with males also demonstrated increases of intra– and interhemispheric coherence between left frontal and posterior brain regions. These coherence increases are presumably correlated with the role of frontal cortex in modulation of posterior cortical regions activity by way of a dynamic filtering or gating mechanism that inhibits irrelevant information (Goldman-Rakis and Friedman, 1991).

Under (ui) instruction, in the theta1 band, females also displayed more cooperation between areas of the left hemisphere than did males. But sex differences were mainly presented in indices of interhemispheric cooperation. When constructing unique images, females seem to need more than males cooperation between frontal parts of the different hemispheres in the theta1 band whereas they display increased cooperation between left frontal and right central and parietal areas in the theta2 band. The increase of interhemispheric frontal cooperation during creative thinking was shown in the number of studies (Bekhtereva et al., 2001; Dan'ko et al., 2003). It is possible that under demand to create unique images females to a greater extent relied on processes related to regulation of attention, planning and foresight. It is known that such functions are linked to frontal lobe integrity.

The above mentioned gender differences in coherence concern theta frequency bands. Increase of theta coherence was primarily associated with memory processes (Sarnthein et al., 1998; von Stein and Sarnthein, 2000; Bhattacharya and Petsche 2002). Thus, one may assume that increased functional connectivity over left hemisphere and over left frontal regions paired with right frontal and right centro-parietal regions in females is associated with recruitment of memory recourses. Speculatively, control of the retrieval and mental scanning of memories containing objects may be involved (Goldman-Rakis and Friedman, 1991; Roland and Gulyas, 1994).

Thus the theta results may suggest that in women-students creative visual thinking is more consciously controlled than in men. However creativity is rather diverse, converting fields such as science and art (Runco, 2005). This may be the reason for differences between our natural sciences faculty students and artistically gifted women who show functional decoupling of the two frontal lobes while mentally creating a picture (Petsche, 1996).

In contrast to changes in theta bands, task-induced changes in both the beta1 (to a lesser extent) and the beta2 frequency bands manifest themselves mainly as more pronounced coherence increases in males in comparison with females. In the beta1, more functional coupling in the left frontal, between the left and right frontal as well as in and between posterior cortical regions was found in males in comparison with females under (i) instruction. The left frontal area below the F3 electrode seems to play a more important role during imagery task performance in males. This effect may be interpreted as increased significance of left frontal lobe functions of expert control and secondary evaluative analysis (Gainotti et al., 1993). The interhemispheric coupling between posterior (temporo-parieto-occipital) areas in beta1 seems to be stronger in males than in females especially under (ui) instruction. This effect may reflect a more intensive informational transfer between both hemispheres in males than in females. The area of posterior parietal cortex is known as a supramodal, integrating area involved in semantic processing (Butters and Brody, 1968; Luria, 1973; Hart and Gordon, 1990). Posterior parietal cortex is also considered as part of the attentional system (Posner et al., 1988; Mesulam, 1990). Temporal cortex, on the other hand, is a higher other association cortex involved in the processing of complex visual forms and object recognition (Desimone et al., 1984). Temporal cortex has also been shown to be involved in visual imagery (Roland, Gulyas, 1994). Using PET and fMRI studies, it was shown that primary visual cortex (occipital regions) may be involved in visual imagery (Sabbab et al., 1995; Kosslyn et al., 1997, 1999; Chen et al, 1998). Von Stein and collaborators (1999) provide evidence that increase of coherent activity in the beta1 range between temporal and parietal cortex exist during higher-level semantic processing of objects. This is in consequence with other studies that show beta frequency to play an important role in higher-level cortical processing (Rescher and Rappelsberger, 1996; 1999; Petche and Etlinger, 1998). Therefore, gender differences in interhemispheric coupling between the abovementioned (temporo-parieto-occipital) areas may indicate an enhanced significance of interactions between hemispheric visual and higher order representations for figural creative thinking in male subjects.

In turn, in the beta2 frequency band in the male group but not in females, diffuse intra- and interhemispheric short- and long–distance coherence increases were found under (i) and (ui) instructions. To some extend, this finding correlates with the results that show beta2 coherence to be higher during successful creative thinking in males as compared to females (Razumnikova, 2004). It is shown that enhanced synchrony in the high frequency beta and gamma bands is most likely due to capabilities for enhanced binding of numerous visual attributes (Bhattacharya and Petsche, 2002). One may assume that the increased functional connectivity over frontal regions is associated with recruitment of attentional recourses whereas increased visual processing, associative and imagery processes are indexed by enhanced posterior functional coupling. Finally, the recruitment of long-distance intra- and interhemispheric connections may be necessary for total increase in all possible association in response to a presented visual fragment. The effects were more pronounced under (i) instruction, when task solutions were not limited by originality requirements.

Thus, it may be concluded that figural creative thinking demonstrates sex-related differences of EEG changes predominantly in theta and beta ranges. We propose that during visual imagery gender specific functional integration is performed by activation and synchronization among different neuronal assemblies evolving in different frequency ranges.

Although the precise mechanisms underlying these differences remain to be specified, the present findings provide further evidence that sex differences may exist in fundamental aspects of functional cortical organization, and not merely in the degrees of expression of invariant principles. Despite the EEG data shows that men and women differ in the neural network engaged during creative figural imagery, they performed the task at the same level. This suggests that different mental strategies may have the same outcomes.

ACKNOWLEDGEMENTS

This research was supported by Russian Fond for Humanity Research, project № 05-06-06234a.

REFERENCES

Abt, K. Descriptive data analysis (DDA) in quantitative EEG studies. In: Samson-Dollfus D, editor. *Statistics and Topography in Quantitative EEG.* Amsterdam: Elsevier; 1988; P.150-160.

Aftanas, LI. Human emotional space: psychophysiological analysis. Novosibirsk: Publ. *SB RAMS;* 2000. (in Russian).

Aftanas, LI.; Lotova, NV; Koshkarov, VI; Popov, SA. Non-linear dynamical coupling between different brain areas during evoked emotions: An EEG investigation. *Biological Psychology,* 1998, V.48, P.121-138.

Annett, MA. A classification of hand preference by association analysis. *Brit. J. Psychol.*, 1970, V.61, P.303-321.

Athley, RA; Keeney, M; Burgess, C. Cerebral hemispheric mechanisms linking ambiguous word meaning retrieval and creativity. *Brain Cogn.,* 1999, V.40, P.479-499.

Babiloni, C; Babiloni, F; Carducci, F; Cappa, SF; Cincotti, F; Del Percio, C; Miniussi, C; Moretti, DV; Rossi, S; Sosta, K; Rossini, PM. Human cortical responses during one-bit short-term memory. A high-resolution EEG study on delayed choice reaction time tasks. *Clin. Neurophysiol,* 2004, V.115, P.161-170.

Bastiaansen, MCM; Posthuma, D; Groot, PFC; de Geus EJC. Event-related alpha and theta responses in a visuo-spatial working memory task. *Clin. Neurophysiol,* 2002, V.113, P.1882–1893.

Bekhtereva, NP; Dan'ko, SG; Starchenko, MG; Pakhomov, SV; Medvedev, SV. Research of brain organization of the creative process. Part III. Brain activation according to data from analysis of local brain blood circulation and EEG. Fiziol Cheloveka, 2001, V.27, P.6-14 (in Russian).

Bhattacharya, J; Petsche, H. Shadows of artistry: cortical synchrony during perception and imagery of visual art. *Brain Res. Cogn. Brain Res.*, 2002, V.13, P.179-186.

Braitenberg, V; Schuz, A. *Anatomy of the Cortex: Statistics and Geometry.* Berlin: Springer-Verlag; 1991.

Butters, N; Brody, BA. The role of the left parietal lobe in the mediation of intra-and cross-modal associations. *Cortex*, 1968, V.4, P.328-343.

Carlsson, I; Wendt, PF; Risberg, J. On the neurobiology of creativity. Differences in frontal activity between high and low creative subjects. *Neuropsychologia,* 2000, V.38, P.873-885.

Chand, I; Runco, MA. Problem finding skills as components in the creative process. *Personality and Individual Differences,* 1993, V.14, P.155-162.

Chen, W; Kato, T; Zhu, XH; Ogawa, S; Tank, DW; Ugurbil, K. Human primary visual cortex and lateral geniculate nucleus activation during visual imagery. *Neuroreport*, 1998, V.9, P.3669-3674.

D'Esposito, M; Detre, JA; Aguirre, GK; Stallcup, M; Alsop, DC; Tippet, L; Farah, MJ. A functional MRI study of mental imagery generation. *Neuropsychologia*, 1997, V. 35, P. 725-730.

Dan'ko, SG; Starchenko, MG; Bekhtereva, NP. EEG local and spatial synchronization during a test on the insite strategy of solving creative verbal tasks. *Fiziol. Cheloveka*, 2003, V.29, P.129-32 (in Russian).

Davidson, R J; Schwartz, GE; Pugash, E; Bromfield, E. Sex differences in patterns of EEG asymmetry. Biol. Psychol., 1976, V.4, P.119-138.

De'Simone, R; Albright, TD; Gross, CG; Bruce, C. Stimulus-selective properties of inferior temporal neurons in the macaque. *J. Neurosci.*, 1984, V.4, P.2051-2062.

DeMoss, K; Milich, R; DeMers, S. Gender, creativity, depression, and attributional style in adolescents with high academic ability. *J. Abnorm. Child. Psychol.*, 1993, V.21, №.4, P.455-467.

Deutsch, G; Bourbon, WT; Papanicolaou, AC; Eisenberg, HM. Visuospatial tasks compared via activation of regional cerebral blood flow. *Neuropsychologia*, 1988, V.26, P.445-452.

Farah, MJ. The neural basis of mental imagery. *Trends Neurosci.,* 1989, V.12. P.395-399.

Gainotti, G; Caltagirone, C; Zoccolotti, P. Left / right and cortical / subcortical dichotomies in the neuropsychological study of human emotions. *Cognition and Emotion,* 1993, V.7, P.71-93.

Goldman-Rakic, PS; Friedman, HR. The circuitry of working memory revealed by anatomy and metabolic imaging. In: Levin HS, Eisenberg HM, Benton AL, editors. *Frontal lobe function and dysfunction.* New York, Oxford: Oxford University Press; 1991; P.72-91.

Gundel, A; Wilson, GF. Topographical changes in the ongoing EEG related to the difficulty of mental tasks. *Brain Topogr.,* 1992, V.5, P.17-25.

Gur, RC; Gur, RE; Obrist, WD; Hungerbuhler, JP; Younkin, D; Rosen, AD; Skolnik, BE; Reivich, M. Sex and handedness differences in cerebral blood flow during rest and cognitive activity. *Science,* 1982, V.217, P.659-661.

Halpern, DF. *Sex differences in cognitive abilities*. 3 rd.ed. Mahwah, NewYork, London: Lawrence Erlbaum Associates; 2000.

Harrington, DM. Effects of explicit instructions to "be creative" on the psychological meaning of DT test scores. *Journal of Personality*, 1975, V.43, P.434-454.

Hart, JJr; Gordon, B. Delineation of single word semantic comprehension deficits in aphasia, with anatomical correlation. *Ann. Neurol.,* 1990, V.27, P.226-231.

Hausmann, M; Gunturkun, O. Sex differences in functional cerebral asymmetries in a repeated measures design. *Brain Cogn.,* 1999, V.41, P.263-275.

Hyde, JS; Linn, MC. Gender differences in verbal ability: A meta-analysis. *Psychological Bulletin,* 1988, V.104, P.53-69.

Inouye, T; Shinosaki, K; Iyama, A; Matsumoto, Y. Localization of activated areas and directional EEG patterns during mental arithmetic. *Electroencephalogr. Clin. Neurophysiol.,* 1993, V.86, P.224-230.

Ivanitskii, AM; Podkletnova, IM; Taratynova, GV. The dynamics of intracortical interaction during thinking activities. *Zh. Vyssh. Nerv. Deiat.,* 1990, V.40, P.230-237 (in Russian).

Kimura, D. Are men's and women's brains really different? *Can. Psychol.,* 1987, V.28, P.133-147.

Klimesch, W; Doppelmayr, M; Schimke, H; Ripper, B. Theta synchronization and alpha desynchronization in a memory task. *Psychophysiology,* 1997, V.34, P.169-178.

Kosslyn, SM; Behrmann, M; Jeannerod, M. The cognitive neuroscience of mental imagery. *Neuropsychologia,* 1995a, V.33, P.1335-1344.

Kosslyn, SM; Chabris, CF; Marsolek, CJ; Koenig, O. Categorical versus coordinate spatial representations: Computational analyses and computer simulations. *J. Exp. Psychol. Hum. Percept. Perform,* 1992, V.18, P.562-577.

Kosslyn, SM; Maljkovic, V; Hamilton, SE; Horwitz, G; Thompson, WL. Two types of image generation: evidence for left and right hemisphere process. *Neuropsychologia,* 1995b, V.33, P.1485-1510.

Kosslyn, SM; Pascual-Leone, A; Felician, O; Camposano, S; Keenan, JP; Thompson, WL; Ganis, G; Sukel, KE; Alpert, NM. The role of area 17 in visual imagery: convergent evidence from PET and rTMS. *Science,* 1999, V.284, P.167-170.

Kosslyn, S.M; Thompson, W.L; Alpert, NM. Neural systems shared by visual imagery and visual perception: A positron emission tomography study. *Neuroimage,* 1997, V. 6, P.320-334.

Lacroix, D; Chaput, Y; Rodriguez, J-P; Filion, M. Quantified EEG changes associated with a positive clinical response to clozapine in schizophrenia. *Prog. Neuropsychopharmacol. Biol. Psychiatry,* 1995, V.19, P.861-876.

Lang, W; Lang, M; Kornhuber, A; Diekmann. V; Kornhuber, HH. Event-related EEG-spectra in a concept-formation task. Hum. *Neurobiol.,* 1988, V.6, P.295-301.

Laukka, SJ; Jarvilehto, T; Alexandrov, YI; Lindqvist, J. Frontal midline theta related to learning in a simulated driving task. *Biol. Psychol.,* 1995, V.40, P.313-320.

Linn, MC; Petersen, AC. A meta-analysis of gender differences in spatial ability: Implications for mathematics and science achievement. In: Hyde JS; Linn MC, editors. *The psychology of gender: Advances through meta-analysis.* Baltimore: Johns Hopkins University Press; 1986; P.67-101.

Linn, MC; Petersen, AC. Emergence and characterization of sex differences in spatial ability: A meta-analysis. *Child Development,* 1985, V.56, P.1479-1498.

Logie, RH. Visuo-spatial working memory. Hillsdale, New York: Lawrence Erlbaum Associates; 1995.

Luria, A.R. *The working brain.* New York. Penguin; 1973.

Maccoby, EE; Jacklin, CN. *The psychology of sex differences*. Stanford, CA: Stanford University Press; 1974.

McIntosh, AR; Grady, CL; Ungerleider, LG; Haxby, JV; Rapoport, SI; Horwitz, B. Network analysis of cortical visual pathways mapped with PET. *Journal of Neuroscience*, 1994, V.14, P.655-666.

Mesulam, MM. Large scale neurocognitive networks and distributed processing for attention, language and memory. *Ann. Neurol.*, 1990, V.28, P.597-613.

Owen, S.V; Baum, S.M. The validity of the measurement of originality. *Educational and Psychological Measurement,* 1985, V.45, P.939-944.

Petsche, H. Approaches to verbal, visual and musical creativity by EEG coherence analysis. *Int. J. Psychophysiol.,* 1996, V.24, P.145-159.

Petsche, H; Etlinger, SC. EEG aspects of cognitive processes: a contribution to the proteus-like nature of consciousness. *Int. J. Psychol.,* 1998, V.33, P.199-212.

Petsche, H; Kaplan, S; von Stein, A; Filz, O. The possible meaning of the upper and lower a frequency ranges for cognitive and creative tasks, *Int. J. Psychophysiol.*, 1997, V.26, P.77–97.

Petsche, H; Pockberger, H; Rappelsberger, P. EEG topography and mental performance. In: Duffy FH, editor. *Topography Mapping of the Brain*. Butterworth: Stoneham; 1986; P.63-98.

Posner, MI; Peterson, SE; Fox, PT; Raichle, ME. Localization of cognitive operations in the human brain. *Science*, 1988, V.240, P.1627-1631.

Razoumnikova, OM. Functional organization of different brain areas during convergent and divergent thinking: an EEG investigation. *Cogn. Brain Res.,* 2000, V.10, P.11-18.

Razumnikova, OM. Gender differences in hemispheric organization during divergent thinking: an EEG investigation in human subjects. *Neurosci. Lett.,* 2004, V.362, P.193-195.

Rescher, B; Rappelsberger, P. Gender dependent EEG – changes during a mental rotation task. *Int. J. of Psychophysiol.,* 1999, V.3, P.209-222.

Roberts, JE; Bell, MA. The effects of age and sex on mental rotation performance, verbal performance, and brain electrical activity. *Dev. Psychobiol.,* 2002, V.40, P.1-18.

Roberts, JE; Bell, MA. Sex Differences on a mental rotation task: variation in electroencephalogram hemispheric activation between children and college students. *Dev. Neuropsychol.,* 2000, V.17, P.199-223.

Roland, PE; Gulyas, B. Visual imagery and visual representation. *Trends in Neuroscience*, 1994, V.17, P.281-287.

Runco, MA. Creativity. *Annu. Rev. Psychol.*, 2004, V.55, P.657-687.

Sabbah, P; Simond, G; Levrier, O; Habib, M; Trabaud, V; Murayama, N; Mazoyer, BM; Briant, JF; Raybaud, C; Salamon, G. Functional magnetic resonance imaging at 1.5 T during sensorimotor and cognitive task. *Eur. Neurol.*, 1995, V.35, P131-136.

Sarnthein, J; Rappelsberger, P; Shaw, GL; von Stein, A. Synchronization between prefrontal and posterior association cortex during working memory tasks in humans, *Proc. Natl. Acad. Sci. USA*, 1998, V.95, P.7092-7096.

Shah, P; Miyake, A. The separability of working memory resources for spatial thinking and language processing: An individual differences approach. *J. Exp. Psychol. Gen.*, 1996, V.125, P.4-27.

Shaywitz, BA; Shaywitz, SE; Pugh, KR; Constable, RT; Skudlarski, P; Fulbright, RK; Bronen, RA; Fletcher, JM; Shankwelier, DP; Katz, L; Gore, JC. Sex differences in the functional organization of the brain for language. *Nature,* 1995, V.373, P.607-609.

Shutiva, CL. Creativity differences between reservation and urban American Indians. *Journal of American Indian Education,* 1991, V.31, P.33-52.

Simonov, PV. Motivated brain. Moscow: *Nauka*; 1987. (in Russian).

Stephens, KR; Karnes, FA; Whorton, J. Gender differences in creativity among American Indian third and fourth grade students. *J. American Indian Education,* 2001, V.40, P.57-65.

Stroganova, TA; Orekhova, EV; Posikera, IN. Externally and internally controlled attention in infants: an EEG study. *Int. J. Psychophysiol.,* 1998, V.30, P.339-351.

Thateher, RW; Krause, PJ; Hrybyk, M. Cortico-cortical association and EEG coherence: a two-compartmental model. *Electroencephalog. Clin. Neurophysiol.*, 1986, V.64, P.123-146.

Torrance, EP. *The Torrance tests of creative thinking streamlined (revised) manual*: Figural A and B. Bensenville, IL: Scholastic testing service; 1984.

Volf, NV; Razumnikova, OM. Sex differences in EEG coherence during a verbal memory task in normal adults. Int. J. Psychophysiol., 1999, V.34, P.113-122.

von Stein, A; Rappelsberger, P; Sarnthein, J; Petsche, H. Synchronization between temporal and parietal cortex during multimodal object processing in man. *Cerebral Cortex*, 1999, V.9, P.137-150.

von Stein, A; Sarnthein, J. Different frequencies for different scales of cortical integration: from local gamma to long range alpha / theta synchronization. *Int. J. Psychophysiol.*, 2000, V.38, P.301-313.

Voyer, D; Voyer, S; Bryden, MP. Magnitude of sex differences in spatial abilities: A meta-analysis and consideration of critical variables. *Psychological Bulletin*, 1995, V.117, P.250-270.

Warren, GH; Luria, Z. Evaluational set and creativity. *Perceptual Motor Skills*, 1972, V.1, P.436-438.

Weiss, S; Rappelsberger, P. EEG coherence within the 13-18 Hz bend as a correlate of a distinct lexical organization of concrete and abstract nouns in humans. *Neurosci. Lett.*, 1996, V.209, P.17-20.

Wendt, PE; Risberg, J. Cortical activation during visual spatial processing: Relation between hemispheric asymmetry of blood flow and performance. *Brain and Cognition,* 1994, V.24, P.87-103.

Zaidel, E; Aboitiz, F; Clark, J. Sexual dimorphism in interhemispheric relations: anatomical-behavioral convergence. *Biol. Res.* 1995, V.28, P.27-43.

In: Focus on Brain Research
Editor: Casey J. Resch, pp. 143-153
ISBN 978-1-60021-395-3

Chapter VI

A Soy-Free Diet Can Lower Testosterone Levels in Male Hamsters: Implications for Aggression Research

Timothy O. Moore[1*] and Mary Karom[2]
[1]Department of Psychology, Clark Atlanta University, Atlanta, GA, USA
[2]Departments of Biology and Psychology, Georgia State University, Atlanta, GA, USA

Abstract

Standard animal lab chow is soy-based, and diets rich in soy protein contain estrogen-like compounds called phytoestrogens. The purpose of this study was to explore the hormonal effects of a soy versus a soy-free diet on testosterone levels in male hamsters. In Experiment #1, thirteen adult male hamsters (110-125 g) were individually-housed, and then divided into a soy and a soy-free group. All hamsters had continuous access to food and water, and they were sacrificed after seven weeks. Blood was collected to measure testosterone levels and the testes were removed to analyze the morphology. There was a significant difference in testosterone levels between the Soy (9.11±1.26 ng/ml) and the Soy-Free group (3.89±0.27 ng/ml)[$t(11)=3.74$; $p<0.005$]. There was no difference in testes weight, but the dimensions were different between the Soy (251.2±6.9 mm2) and the Soy-Free group (229.5±6.8 mm2)[$t(13)=2.2$; $p<0.05$]. In Experiment #2, twenty adult male hamsters (110-125 g) were individually-housed and then divided into four groups: 1) Castrate+Soy; 2) Castrate+Soy-Free; 3) Sham Castrate+Soy; and 4) Sham Castrate+Soy-Free. Animals were sacrificed seven weeks after surgery and the diet manipulation. There were no detectable levels of testosterone in castrated males. However, the Sham+Soy (12.21±2.95 ng/ml) had significantly higher levels of testosterone compared to Sham+Soy-Free (3.92±0.8 ng/ml)[$t(9)=2.95$; $p=0.01$]. In sum, we have demonstrated that a soy-free diet lowered endogenous levels of

* Correspondence should be addressed to Timothy O. Moore, Ph. D., Department of Psychology, Clark Atlanta University, 223 James P. Brawley Drive, Atlanta, GA 30314. Tel: 404-880-6939; Fax: 770-879-0275; Email: tmoore@cau.edu.

testosterone, and we recommend that investigators evaluate the impact of phytoestrogens when investigating behavioral experiments that involve neurosteroids.

INTRODUCTION

Interpreting the results of experiments that are designed to investigate hormones and social behavior can be a challenge (Albers, Huhman and Meisel, 2002). Although researchers may make every attempt to design properly controlled experiments, there may be unseen factors that can affect social behavior. Hamsters are an excellent animal model to study interactions between conspecifics in a social setting. Although studies with hamsters can be controlled, diet is one particular extraneous variable that can affect experimental results. In the present study, the implication of diet and hormonal status will be addressed without the context of social behavior. The focal point of this investigation will be on soy products that contain estrogen-like compounds.

Estrogen-like compounds called phytoestrogens are found in soy-based animal diets. Soy products are used as the major protein source in all natural-ingredient, commercially available, rodent diets. The different types of phytoestrogens are isoflavones, coumestans and lignans. In soy-based animal diets, genistein and daidzein are specific isoflavones that can exert estrogen-like effects on behavior.

A recent literature review of hamster aggression and diet revealed less than four experiments that have been reported. The effects of leptin on the estrous cycle (Schneider, Goldman, Tang, Bean, Ji and Friedman, 1998) and the effects of bacteria against neoplasia (Lemonde, 1973) were conducted. However, our study investigating the neurobehavioral effects of phytoestrogens in male hamsters (Moore, Karom and O'Farrell, 2004) was the most recent publication to study the connection between diet and agonistic behavior in male hamsters.

We have consistently demonstrated that testosterone levels are lowered in hamsters fed a soy-free diet. This is a significant finding because testosterone is known to play a critical role in male aggression. If diet can manipulate endogenous levels of testosterone, it is pertinent that diet should be taken into consideration when conducting behavioral experiments.

Testosterone can be converted into estradiol by the enzyme aromatase. In addition, testosterone can be synthesized into 5-DHT by 5-alpha reductase. Therefore, the effects of testosterone may be exerted by either testosterone, 5-DHT or estradiol. Furthermore, the receptors for both testosterone and estrogen may influence agonistic behavior. There is abundant evidence from rodent experiments that testosterone can enhance aggressive behavior. The positive relationship has been primarily observed in mice and rats. In hamsters, this direct relationship has been inclusive (Moore, 2001).

For example, Syrian hamsters are spontaneously aggressive animals that have been used by numerous investigators to study the neural and hormonal control of aggression. Over the last three decades, studies examining the effects of gonadal hormones on intermale aggression in hamsters have produced conflicting results (for review see Albers et al., 2002). Some studies report that castration and/or administration of gonadal hormones to castrates can significantly influence the expression of at least some measure of aggression (Drickamer,

Vandenbergh and Colby, 1973; Grelk, Papason, Cole and Rowe, 1974, Payne, 1973; Payne, 1974; Payne and Swanson, 1971; Payne, 1972; Potegal, Blau, Black and Glusman, 1980; Vandenbergh, 1971), whereas other studies found no reduction in aggression following castration (Evans and Brain, 1974; Tiefer, 1970; Whitsett, 1975). The differences in the results of these studies may be due to the many different, and sometimes complicated, methodologies that have been used to study the effects of gonadal hormones on aggression in male hamsters.

In retrospect, seldom was there any mention that diet could be a contributing factor in hormone and aggression experiments. In most rodent experiments, the standard animal lab chow is soy-based. Diets rich in soy protein contain called phytoestrogens, and these compounds may indirectly alter physiology and behavior. Before we further investigate the behavioral effects of phytoestrogens, it is important that we understand the general physiological effects of phytoestrogens. In this study, we will explore the hormonal effects of a soy versus a soy-free diet on testosterone levels in male hamsters. The purpose of this investigation will be to explore if diet alone, and not in conjunction with agonistic behavior, can affect the hormonal status of male hamsters.

METHOD

Animals

Adult male hamsters (110-125 g) were purchased from Charles River Laboratory and housed in a light-dark cycle consisting of 14-h light and 10-h dark (lights off at 1000). Food and water were continuously available throughout all experiments. Upon arrival to our laboratory, experimental animals were isolated.

Diet Manipulation

Hamsters were divided into two groups: 1) the standard soy-based 5001 animal diet (Soy); and 2) the 5K96 casein-based diet (Soy-Free). The soy diet contained 460 ppm total isoflavones and the soy-free diet contained less than 1 ppm total isoflavones. The diets were purchased from LabDiet (Purina Mills). The casein-based diet is a natural ingredient diet, formulated to be used in experimental protocols where dietary estrogenic activity is a concern. It consistently analyzes less than 1.0 ppm total isoflavones (aglycone equivalents of genistein, daidzein and glycitein). The Purina Mills 5001 diet contains an equivalent of 810 μg/g total phytoestrogens (Brown and Setchell, 2001).

Experiment #1

Thirteen male hamsters were isolated and then divided into two groups: Soy and Soy-Free. Animals were sacrificed seven weeks after the diet manipulation.

Experiment #2

Twenty male hamsters were isolated and then divided into four groups: 1) Castrate+Soy; 2) Castrate+Soy-Free; 3) Sham Castrate+Soy; and 4) Sham Castrate+Soy-Free. Animals were sacrificed seven weeks after surgery and the diet manipulation.

Surgery

Hamsters were anesthetized in an enclosed Plexiglas chamber filled with isoflurane. Hamsters were shaved and the surgical area was cleaned with isopropropyl alcohol and betadine solution. A face mask was placed over the nasal region and isoflurane gas was continuously delivered for the entire surgical procedure. Hamsters were bilaterally castrated through a midline abdominal incision. After locating and retracting each testicle through the incision, the cremaster muscle was separated from the tail of the epididymis in order to maximally exteriorize the testicle and associated tissues. The ductus deferens, testicular vessels and associated fat were then twisted, ligated and crushed with hemostats distal to the ligature. The testicles were then removed and the stump was allowed to slip back into the abdomen. Neither the testicle nor the surrounding tissue was removed in sham castrates. The abdominal muscle was sutured in a simple continuous pattern and skin incisions were closed with wound clips. The suture material used for both the abdominal closure and the ligations was 4-0 polyglactin 90. Hamsters recovered from anesthesia in a heated room.

Radioimmunoassay

Using methods described elswhere (Albers and Prishkolnik, 1992), testosterone (T) levels were measured by radioimmunoassay (RIA) in blood samples obtained from the inferior vena cava after lethal injection of sodium pentobarbital. Briefly, 5 μl of serum from each sample was dissolved in 196 μl of 0.1 M phosphate-buffered saline (PBS) and extracted in 2 ml of ether. After extraction, each sample was assayed in duplicate, 50 μl antibody (ICN Biomedical, Costa Mesa, CA), and 50 μl of ^{125}I-labeled T (1800cpm), and was allowed to incubate overnight at room temperature. The reported cross-reactivities were as follows: T, 100%; DHT, 3.4%; androstenedione, 0.6%; androsterone, <0.01%, 17β-estradiol, <0.01%; estrone, <0.01%; progesterone, <0.01%. A dextran-coated charcoal incubation (20 min) removed unbound tracers. After centrifugation for 15 min at 4 degrees celcius, the supernatant was decanted and counted by an LKB1272 gamma counter. The intra-assay variability was approximately 10%.

Statistical Analyses

A Student's t-Test for independent samples was used to determine the difference between treatment groups. Data with more than two groups were analyzed by between-groups, one factor analysis of variance (ANOVA). When a significant effect was encountered, a Tukey HSD post hoc test was employed.

RESULTS

Experiment #1

All hamsters were sacrificed seven weeks after the diet manipulation, and blood serum and testes were collected. Fig. 1 illustrates a significant difference in testosterone levels between the Soy (9.11±1.26 ng/ml) and the Soy-Free group (3.89±0.27 ng/ml)[$t(11)=3.74; p<0.005$]. Although the production of testosterone was affected, an analysis of the testes provided intriguing results. There were no differences in testes weights, but the dimensions of the testes were different between the Soy (251.2±6.9 mm^2) versus the Soy-Free group (229.5±6.8 mm^2) [$t(13)=2.2$; $p<0.05$]. Fig. 2 represents the testes dimensions.

As stated, there was no difference in testes weight between groups and there was also no difference in the overall body weight at the time of sacrifice (see Table 1). All hamsters had continuous access to food and water. Therefore, the diet was mainly controlled by the normal consumption of food by each animal. The amount of food consumed was not measured, but there was no difference in body weight between groups. Whether the diet was soy- or casein-based, there was equal growth between groups. However, endogenous hormones such as testosterone appeared to be sensitive to the diet manipulation.

Table 1. Comparison of body and testes weight when animals were sacrificed seven weeks post-diet manipulation.

	SOY	SOY FREE
Body Weight	132.6±2.7	134.3±4.9
Testes Weight	3.3±0.01	3.4+0.0

Exp.#1 Testosterone Levels in Blood Serum

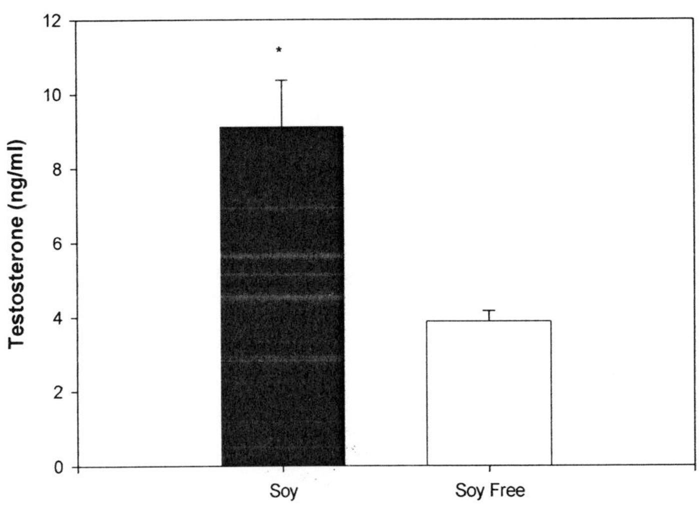

Figure 1. Following the administration of a soy (n=7) or a soy-free (n=6) diet for seven weeks, hamsters were sacrificed and testosterone was measured. An asterisk denotes a significant difference ($p<0.005$). Testosterone was higher in the Soy versus the Soy-Free group.

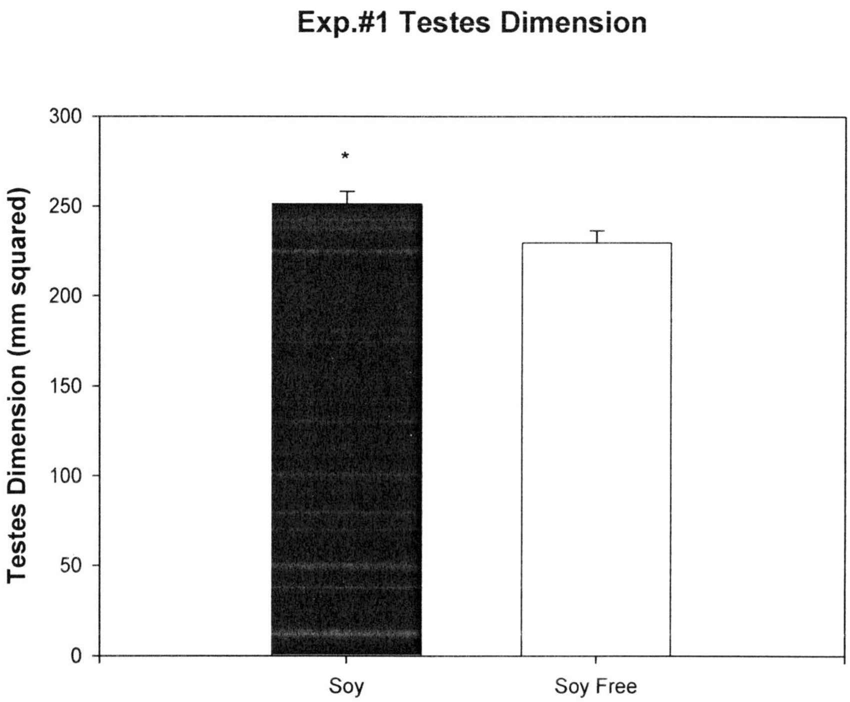

Figure 2. Following the administration of a soy (n=7) or a soy-free (n=6) diet for seven weeks, hamsters were sacrificed and the testes were removed and measured to analyze the size, weight and dimensions. An asterisk denotes a significant difference ($p<0.05$). The dimensions were slightly higher in the Soy group.

Experiment #2

An analysis was conducted between four different groups: Castrate+Soy; Castrate+Soy Free; Sham Castrate+Soy; and Sham Castrate+Soy-Free. Not surprisingly, there were no detectable levels of testosterone in castrated hamsters. According to our results, there was no evidence of any circulating testosterone in the blood serum of castrated hamsters. In contrast, the Sham Castrate+Soy group (12.21±2.95 ng/ml) had significantly higher levels of testosterone compared to the Sham Castrate+Soy-Free group (3.92±0.8 ng/ml) [$t(9)=2.95$; $p=0.01$]. The data are illustrated in Fig. 3

Since one half of the animals were castrated, we did not analyze the testes in Experiment #2. The weight was analyzed in Fig. 4 and we did observe the expectant increase in size over the seven week period for castrated hamsters. Although there were comparable weights between all four groups before and after the diet manipulation, we did observe a percent change within groups. For example, the Castrate+Soy group increased from 117g to 147g ($p<0.009$) and the Castrate+Soy-Free group increased from 123g to 153g ($p<0.09$). The Sham Castrate+Soy group increased from 120g to 132g ($p<0.002$), but the Sham Castrate+Soy-Free group did not have a significant increase from 125g to 153g ($p=0.13$) due to a high standard error of the mean.

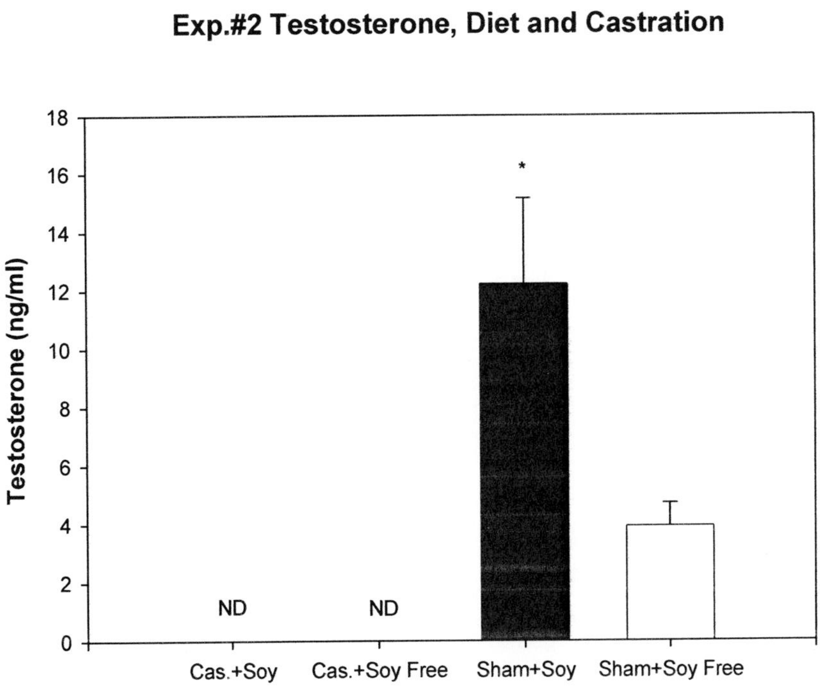

Figure 3. Twenty male hamsters were evenly divided into four groups. Hamsters were sacrificed seven weeks later and testosterone levels were measured. An asterisk denotes a significant difference (p=0.01) between the Sham+Soy and the Sham+Soy Free group. There were no detectable levels in castrates.

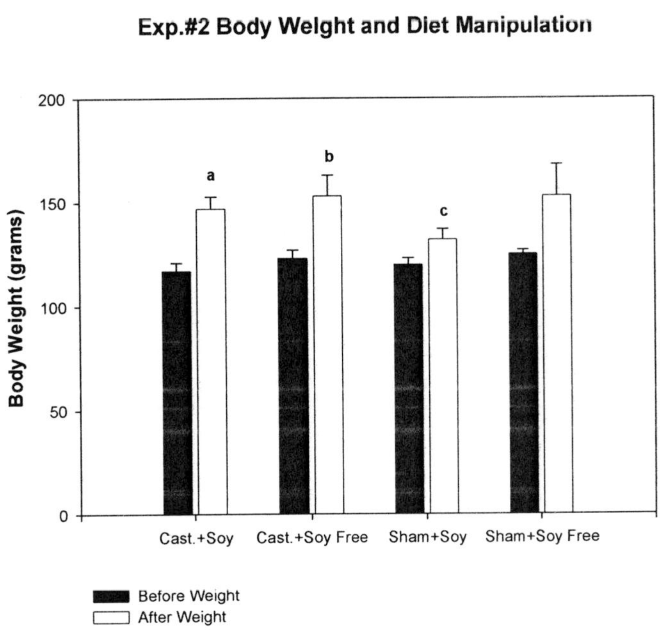

Figure 4. Body weight was measured before and after the diet and surgical manipulation. There was a significant change in weight for three out of the four groups. The letters (a, b and c) denote significance.

CONCLUSION

Our findings demonstrate that testosterone levels in male hamsters can be influenced by a standard, soy-based lab chow. The soy-free diet we used was casein-based, and we have consistently demonstrated that a soy-free diet can lower testosterone levels in the blood serum when compared to a diet containing soy. We did not measure consumption of the diet for each hamster, however, there was no difference between the weight of hamsters fed the different diets. The different levels of testosterone are more than likely due to the content of the food and not the amount of food consumed.

In hamsters, the only other published reports describing the influence of diet on physiology and social behavior have been with female Syrian hamsters fed leptin to study the effects on estrous cyclicity, social behaviors and ovulation rate (Schneider et al., 1998). Moreover, a very old study on the effects of bacteria against neoplasia in mice and hamsters was conducted (Lemonde, 1973). Our recent publication (Moore et al., 2004) has driven our research into uncharted territory. We can reliably suggest that diet should be taken into consideration when studying hormones and social behavior. In the analysis of behavior, the behavior may be indirectly altered by the diet and not the manipulated independent variable (e.g., lighting, opponent, phase, experience, surgical manipulation, etc.).

We have previously demonstrated (Moore et al. 2004) that the Soy-Free group had higher levels of vasopressin V1A receptor binding compared to the Soy group in the lateral septum. In contrast, the soy group had higher receptor binding in the lateral septum when compared to the soy-free group. We observed hormonal changes as a result of the diet manipulation and other research (Davis, 2002) investigating neuroanatomy suggests that soy products or phytoestrogens may be potent selective estrogen receptor modulators (SERMs). The summarized data are not conclusive, but there is speculation that phytoestrogens may exert both estrogenic and anti- estrogenic effects on estrogen receptors α and β. We did not measure any receptors in this study, but we can speculate that soy is affecting more than just blood serum concentrations of testosterone.

Interestingly, there was no difference in the weight of the testes between groups, but the dimensions were slightly different. In other words, the morphology of the testes may have changed from the diet manipulation, but the size was not affected. The change in morphology may have a connection to diet and cancer formation. For example, there has been an interest in potentially anticarcinogenic actions of isoflavones. Laboratory rats fed soy-enriched diets have less breast tumor development following stimulation with the directly- and indirectly-acting tumour-inducers (Messina, 1994).

Some investigators have directly exposed animals to specific phytoestrogens such as genistein and coumestrol. In a recent study, Wisniewski, Cernetich, Gearhart and Klein (2005) determined if exposure to genistein, during early periods of sex differentiation alters reproductive development and behavior in male mice. Aggressive behaviors were decreased, whereas defensive behaviors were increased in males that received the low dose genistein diet. According to Wisniewski et al.(2005), exposure to genistein during critical periods of sex differentiation results in concurrent and persistent demasculinization in male mice. In another study, the effects of coumestrol on estrogen receptor α knockout mice (ErαKO) was investigated (Jacob, Temple, Patisaul, Young and Rissman, 2001). The results demonstrated that coumestrol has antiestrogenic actions in the brain and pituitary and that ERα mediates these effects.

Other central nervous system mechanisms have also been studied. Androgen metabolizing enzymes, for example, are known to regulate neuroendocrine functions and reproductive behaviors, while calcium-binding proteins are associated with protecting against neurodegenerative diseases (Lephart, Thompson, Setchell, Adlercreutz and Weber, 2000). It was reported that consumption of phytoestrogens in rats via a soy diet for a relatively short interval can elevate plasma and brain phytoestrogen levels and decrease brain calcium-binding proteins without altering brain androgen metabolizing enzymes.

Lephart, Galindo and Bu (2001) were the first to show that life-long consumption of dietary phytoestrogens alters hypothalamic-pituitary-adrenal stress response in male rats. Phytoestrogen fed rats displayed significantly higher plasma ACTH and hippocampal glucocorticoid receptor levels when compared to phytoestrogen-free values. Similar to what we have presented in this study, the research by Lephart et al. (2002) provides critical evidence that soy-derived phytoestrogens have complex and beneficial effects on brain structure, function and behavior.

In sum, the neurobehavioral effects of dietary soy phytoestrogens cannot be overlooked in animal research. We must determine the importance of phytoestrogens given the recognized significance of estrogen in brain and neural disorders (Lephart et al., 2002). There is epidemiological, laboratory, and clinical evidence which indicates that phytoestrogens, like certain SERMs, have an antiproliferative effect on the breast, and positive effects on the lipoprotein profile and bone density (Brzezinski and Debi, 1999). If animal studies are to be the baseline for effects on humans, accurate assessment of the data will require researchers to consider the type of diet chosen for experimentation. In conclusion, there are potential benefits to human consumption of a soy-based diet. The research is not conclusive, but well controlled animal experiments will help to determine the beneficial as well as detrimental effects of phytoestrogens on human health.

Acknowledgments

In conjunction with the Center for Behavioral Neuroscience (NSF-STC #IBN-9876754), this work was supported by NIH Grant #NS37232.

References

Albers, H.E., and Prishkolnik, J. (1992). Sex differences in odor stimulated flank marking behavior in the Golden hamster (*Mesocricetus auratus*), *Hormones and Behavior*, 26, 229-239.

Albers, H.E., Huhman, K.L. and Meisel, R.L. (2002). *Hormonal basis of social conflict and communication. Hormones and Behavior*, 1, 393-433. Academic Press, New York.

Brown, N.M.and Setchell, K.D. (2001). Animal models impacted by phytoestrogens in commercial chow. *Laboratory Investigation*, 81, 735-747.

Brzezinski, A. and Debi, A. (1999). Phytoestrogens: the "natural" selective estrogen receptor modulators? *European Journal of Obstetrics and Gynecology and Reproductive Biology*, 85, 47-51.

Drickamer, L.C., Vandenbergh, J.G., and Colby, D.R. (1973). Predictors of dominance in the male golden hamster (*Mesocricetus auratus*). *Animal Behavior*, 21, 557-563.

Davis, S.R., (2002). Phytoestrogens in the context of SERMs. In *Contemporary Endocrinology: Selective Estrogen Receptor Modulators: Research and Clinical Applications* (Eds. Manni, A. and Verderame, M.F.) pp. 345-363, Humana Press, Totowa, NJ.

Evans, C.M. and Brain, P.F. (1974). Some studies on endocrine influences on aggressive behavior in the golden hamster (*Mesocricetus auratus* Waterhouse). *Progress in Brain Research*, 41, 473-480.

Grelk, D.F., Papson, B.A., Cole, J.E. and Rowe, F.A. (1974). The influence of caging conditions and hormone treatments on fighting in male and female hamsters. *Hormones and Behavior*, 5, 355-366.

Jacob, D.A., Temple, J.L., Patisaul, H.B., Young, L.J., and Rissman, E.F. (2001). Coumestrol antagonizes neuroendocrine actions of estrogen via the estrogen receptor α. *Experimental Biology and Medicine*, 226(4), 301-306.

Lemonde, P. (1973). Protective effects of BCG and other bacteria against neoplasia in mice and hamsters. *National Cancer Institute Monograph*, 39, 21-32.

Lephart, E.D., Thompson, J.M., Setchell, K.D.R., Adlercreutz, H. and Weber, K.S. (2000). Phytoestrogens decrease brain calcium-binding proteins but do not alter hypothalamic androgen metablozing enzymes in adult male rats. *Brain Research*, 859, 123-131.

Lephart, E.D., West, T.W., Weber, K.S., Rhees, R.W., Setchell, K.D.R., Adlercreutz, H. and Lund, T.D. (2002). Neurobehavioral effects of dietary soy phytoestrogens. *Neurotoxicology and Teratology*, 24, 5-16.

Lephart, E.D., Galindo, E. and Bu, L.H. (2003). Stress (hypothalamic-pituitary-adrenal axis) and pain response in male rats exposed lifelong to high vs. low phytoestrogen diets. *Neuroscience Letters*, 342, 65-68.

Messina, M. (1994). Soy intake and cancer risk: a review of the in vitro and in vivo data. *Nutrition and Cancer*, 21, 113-131.

Moore, T.O. (2001). Testosterone and male behavior: Empirical research with hamsters does not support the use of castration to deter human sexual aggression. *North American Journal of Psychology*, 3(3), 503-520.

Moore, T. O., Karom, M. and O'Farrell, L. (2004). The neurobehavioral effects of phytoestrogens in male Syrian hamsters. *Brain Research*, 1016, 102-110.

Payne, A.P. and Swanson, H.H. (1971). The effect of castration and ovarian implantation on aggressive behaviour of male hamsters. *Journal of Endocrinology*, 51, 217-218.

Payne, A.P. and Swanson, H.H. (1972). The effect of sex hormones on the agonistic behavior of the male golden hamster (*mesocricetus auratus* waterhouse). *Physiology and Behavior*, 8, 687-691.

Payne, A.P. (1973). A comparison of the aggressive behaviour of isolated intact and castrated male golden hamsters toward intruders introduced into the home cage. *Physiology and Behavior*, 10, 629-631.

Payne, A.P. (1974). A comparison of the effects of androstenedione, dihydrotestosterone and testosterone propionate on aggression in the castrated male golden hamster. *Physiology and Behavior*, 13, 21-26.

Potegal, M., Blau, A.D., Black, M. and Glusman, M. (1980). Effects of castration of male golden hamsters on their aggression toward a restrained target. *Behavioral Neural Biology*, 29(3), 315-330.

Schneider, J.E., Goldman, M.D., Tang, S., Bean, B., Ji, H. and Friedman, M.I. (1998). Leptin indirectly affects estrous cycles by increasing metabolic fuel oxidation. *Hormones and Behavior*, 33(3), 217-228.

Tiefer, L. (1970). Gonadal hormones and mating behavior in the adult golden hamster. *Hormones and Behavior*, 1, 189-202.

Vandenbergh, J.G. (1971). The effects of gonadal hormones on the aggressive behavior of adult golden hamsters (*Mesocricetus auratus*). *Animal Behavior*, 19, 589-594.

Whitsett, J.M. (1975). The development of aggressive and marking behavior in intact and castrated male hamsters. *Hormones and Behavior*, 6, 47-57.

Wisniewski, A.B., Cernetich, A., Gearhart and Klein, S.L. (2005). Perinatal exposure to genistein alters reproductive development and aggressive behavior in male mice, 84, 327-334.

INDEX

A

absorption, 99
access, x, 143, 147
accounting, 38
acetylcholine, 39, 55, 78, 83
acetylcholinesterase, 78
acetylcholinesterase inhibitor, 78
achievement, 139
acid, 2, 4, 8, 10, 11, 20, 22, 23, 28, 31, 32, 38, 45, 92, 94, 96, 97, 98, 102, 103, 105
ACTH, 151
activation, viii, 11, 12, 13, 14, 15, 16, 21, 23, 25, 28, 30, 32, 35, 38, 40, 41, 42, 43, 44, 45, 46, 47, 48, 49, 50, 51, 52, 53, 54, 55, 56, 57, 58, 59, 60, 62, 69, 70, 71, 72, 73, 75, 77, 78, 79, 80, 81, 83, 86, 110, 111, 112, 113, 114, 115, 116, 117, 118, 119, 121, 137, 138, 140, 141
activation state, 60
active transport, 98
adaptation, 41
adenine, 7, 9
adenosine, 7, 10, 18, 22, 81
adenosine triphosphate, 7, 10, 18, 22
ADHD, 110
adhesion, 41, 55, 56, 59, 61, 67, 70
adolescents, 133, 138
ADP, 21
adults, 141
age, 30, 40, 41, 42, 53, 57, 63, 67, 76, 93, 95, 133, 140
ageing, 64
agent, 20, 101
aggregates, 22, 36, 37, 59, 77
aggregation, viii, 35, 40, 45, 48, 55, 59, 60, 77
aggression, 144, 145, 152, 153
aggressive behavior, 144, 152, 153
aging, vii, 17, 35, 36, 37, 39, 40, 41, 42, 43, 48, 52, 53, 58, 59, 60, 63, 69, 74, 76, 77, 81, 83, 84, 96
aging process, 69
AIDS, 93
alcohol, 146
alcoholism, 93
aldehydes, 4
allele, 19, 21, 38, 75
alternative, 38, 39, 50, 109
alters, 30, 74, 150, 151, 153
aluminum, 92
American Indian, 141
American Indians, 141
amino acid, 8, 9, 10, 11, 22, 28, 38, 47, 84
amino acids, 8, 9, 10, 11, 28, 38, 47
ammonium, 102
amoeboid, 68, 73
amplitude, 17, 55, 125
amyloid beta, 26, 62, 63, 64, 67, 68, 69, 71, 75, 76, 77, 78, 79, 80, 83, 84, 85, 86
amyloid deposits, 46, 74
amyloid plaques, 36, 37, 40, 43, 45, 46, 49, 56, 58, 75, 76, 82, 93
amyloid-β peptide, 33
amyloidosis, 74, 82
anatomy, 138
androgen, 151, 152
androstenedione, 146, 153
animal models, vii, 1, 25, 52
animals, 43, 49, 91, 144, 145, 147, 148, 150
anion, 2, 3, 92, 96, 99, 100, 101, 103, 104
ANOVA, 146
anoxia, 28

antagonism, 32
antibiotic, 103
antibiotics, 93, 94
antibodies, 94
antibody, 51, 103, 104, 146
antibody (Ab), 104
anticancer drug, 104
anticonvulsants, 93
antidepressant, 39, 78
antigen, 78, 81
anti-inflammatory drugs, 44, 45, 79
antioxidant, vii, 1, 3, 4, 6, 9, 11, 13, 14, 15, 16, 18, 20, 24, 25, 31, 32, 38, 41, 45, 51, 61, 75, 81, 86
antisense, 94
antisense oligonucleotides, 94
anxiety, 17
aorta, 97
apathy, 17
aphasia, 120, 138
apical, 99
apoptosis, 5, 11, 12, 18, 21, 22, 24, 27, 29, 31, 32, 53, 64, 66, 74, 81, 82, 83, 84
apoptotic pathway, 26
arachidonic acid, 16
artery, 32, 85
ascorbic acid, 2, 3
aspartate, 77
assessment, 100, 109, 112, 119, 120, 121, 151
astrocytes, 6, 7, 8, 9, 10, 13, 14, 15, 16, 24, 25, 28, 29, 33, 36, 38, 43, 44, 45, 46, 47, 48, 49, 50, 51, 53, 55, 56, 57, 59, 60, 61, 63, 64, 67, 70, 71, 75, 76, 77, 78, 79, 80, 81, 82, 83, 84, 91, 98, 101
asymmetry, 19, 108, 110, 111, 120, 134, 138
ATP, 7, 9, 10, 11, 15, 16, 17, 18, 19, 21, 22, 24, 28, 29, 81, 103
attention, ix, 107, 110, 112, 114, 116, 120, 134, 135, 140, 141
autism, ix, 107, 110
autoimmune disease, 52
autoimmune diseases, 52
autopsy, 67
availability, 8, 10
averaging, 117, 127
axons, 6, 15, 16

B

bacteria, 44, 77, 144, 150, 152
bacterial, 94
barriers, 98
basic fibroblast growth factor, 2, 16
BBB, viii, 87, 90, 91, 92, 93, 94, 95
BD, 98, 103, 119
behavior, 111, 144, 145, 150, 151, 152, 153
behavioral effects, 145
beneficial effect, 43, 52, 53, 60, 151
betadine, 146
bias, 78, 118
bilirubin, vii, 1, 6, 24, 27, 33
binding, 7, 13, 22, 27, 32, 55, 56, 57, 60, 62, 68, 84, 103, 136, 150, 151, 152
biological activity, 83
biology, 91
biosynthesis, 27
bladder, 17
blocks, 39, 44
blood, viii, 21, 40, 87, 88, 89, 90, 92, 95, 96, 97, 98, 99, 100, 101, 102, 103, 104, 109, 137, 141, 146, 147, 148, 150
blood flow, 21, 40, 141
blood vessels, 97
blood-brain barrier, viii, 88, 90, 92, 93, 94, 95, 96, 97, 98, 100, 101, 102, 103, 104
blood-brain barrier (BBB), viii, 87, 88, 90, 91, 92, 93, 94, 95, 96, 97, 98, 100, 101, 102, 103, 104
body weight, 5, 147
bonds, 3
boys, 120
brain activity, 109, 113, 114, 117
brain asymmetry, 110, 124
brain damage, 7, 21, 25, 40, 85, 109
brain functions, ix, 107, 109, 111, 120
brain lateralization, 119
brain structure, 151
brain tumor, 93, 100
by-products, 16, 92

C

Ca^{2+}, 11, 12, 13, 14, 16, 19, 21, 22, 24, 25
calcium, 12, 13, 22, 25, 39, 40, 53, 61, 64, 65, 77, 151, 152
cancer, 90, 92, 100, 105, 150, 152
cancer cells, 90, 100
candidates, 109
capillaries, viii, 87, 88, 91, 97, 99, 103, 104
capillary, 89, 90, 91, 92, 96, 99, 100, 104
carbohydrate, 42
carbohydrate metabolism, 42
cardiac, 101

cardiac glycoside, 101
carrier, 103
casein, 145, 147, 150
caspases, 11, 12, 16
castration, 144, 152, 153
catabolism, 24
catalase, 2, 3, 4, 29, 51
cation, 97, 99, 100, 103, 105
CD14, 72
cDNA, 30, 103
cell, vii, 1, 3, 5, 6, 7, 8, 9, 10, 11, 12, 13, 14, 15, 16, 17, 18, 19, 21, 22, 23, 24, 25, 28, 29, 30, 32, 33, 35, 37, 38, 39, 40, 41, 42, 44, 46, 47, 48, 49, 50, 51, 52, 53, 54, 55, 57, 58, 59, 62, 63, 64, 66, 70, 75, 81, 82, 83, 85, 92, 96, 97, 100, 105
cell adhesion, 37
cell body, 21
cell culture, 32, 47, 53, 66
cell cycle, 5, 12, 13, 25, 29, 44
cell death, vii, 1, 5, 9, 11, 12, 15, 18, 19, 21, 22, 23, 24, 25, 28, 29, 30, 33, 35, 48, 50, 52, 64, 70, 81, 82, 85
cell growth, 55
cell line, 32, 75, 92, 97, 100, 105
cell lines, 75, 92, 100, 105
cell membranes, 5, 55
cell signaling, 42
cell surface, 70
cells, viii, 87, 88, 89, 91, 99, 101, 103
central nervous system, vii, viii, 1, 27, 35, 53, 62, 63, 64, 66, 72, 76, 96, 98, 100, 101, 103, 151
central nervous system (CNS), viii, 87, 89, 90, 92, 93, 95, 96, 97, 98, 100, 101, 102, 103, 104
cephalosporin, 94
cephalosporin antibiotics, 94
cerebral blood flow, 109, 138
cerebral cortex, 52, 61, 74, 83
cerebral hemisphere, 114
cerebral ischemia, 31, 32, 82, 83
cerebrospinal fluid, 59, 89, 90, 97, 98, 99, 102, 103, 104, 105
cerebrospinal fluid (CSF), 89, 90, 91, 93, 97, 98, 99, 101, 102, 103, 104, 105
certainty, 93
channels, 11
chelates, 20
chemiluminescence, 81
chemokines, 48, 50, 51, 56, 57, 71
chemotaxis, 78, 81
chemotherapeutic, 92
chemotherapy, 93
children, 133, 140
chlorine, 21
cholesterol, 31, 40, 41, 43, 55, 75, 84
cholesterol-lowering drugs, 41
cholinesterase, 43
choroid, viii, 87, 88, 90, 91, 92, 95, 96, 97, 101, 102, 104
circulation, 90, 137
cisterna magna, 90
classification, 137
cleavage, 12, 19, 40, 45, 58, 63
clinical trials, vii, 1, 25
cloning, 93, 99, 100, 103
closure, 146
clozapine, 139
cluster of differentiation (CD), 97
clusters, 115
CNS, viii, 39, 41, 46, 48, 50, 51, 52, 57, 63, 68, 73, 81, 83, 87, 89, 90, 92, 93, 95, 97, 102, 104
coenzyme, 31, 84
cognition, 19, 40, 85, 111
cognitive abilities, 109, 124, 138
cognitive activity, 138
cognitive deficit, 39, 55, 77, 81
cognitive deficits, 55, 77
cognitive function, ix, 43, 61, 107, 108, 109, 110, 111, 112
cognitive impairment, 37
cognitive performance, 31
cognitive process, 43, 111, 140
cognitive tasks, 109, 111
coherence, ix, 123, 125, 127, 129, 130, 131, 132, 133, 134, 135, 136, 140, 141
cohort, 117
college students, 140
communication, 46, 108, 151
communities, 95
community, 78
competition, 8, 57
complement, 50, 57, 59, 74, 84
complementary DNA (cDNA), 103
complex behaviors, 51
complexity, vii, 6, 11, 95
complications, 109
components, viii, 75, 77, 87, 111, 124, 127, 138
compounds, vii, viii, x, 3, 8, 18, 20, 27, 35, 36, 48, 58, 59, 87, 90, 91, 92, 93, 94, 95, 102, 143, 144, 145
comprehension, vii, 1, 25, 138

computer simulations, 139
concentration, 9, 10, 17, 33, 39, 47, 57, 134
concrete, 141
conduction, 6
configuration, vii, 1, 25
conflict, 151
conjugation, 94
connectivity, 109, 135, 136
consciousness, 140
constipation, 17
construction, 135
consumption, 7, 9, 147, 150, 151
context, 144, 152
control, ix, 9, 48, 57, 73, 76, 79, 107, 116, 135, 136, 144
control condition, 48
convergence, 141
conversion, 23
copper, 3, 4, 19, 23, 32
correlation, viii, 17, 65, 79, 80, 87, 138
correlations, 111
cortex, 36, 55, 65, 118, 119, 121, 136, 140
cortical neurons, 69
cortical processing, 136
coupling, ix, 6, 8, 10, 111, 123, 132, 133, 136, 137
creatine, 19, 21, 31
creative process, 137, 138
creative thinking, 125, 129, 135, 136, 137, 141
creativity, ix, 123, 124, 126, 128, 135, 137, 138, 140, 141
Crigler-Najjar syndrome, 33
critical period, 150
critical variables, 141
CRP, 92
CSF, 48, 57, 68, 72, 73, 82, 89, 90, 91, 93, 98, 101
CTA, 114, 115
culture, 15, 28, 29, 45, 47, 50, 56, 59, 64, 69, 80, 82, 97
cultured, 91, 99, 103, 104
cultures, 91, 96, 98, 100
curing, 25
cycles, 153
cyclooxygenase, 2, 16, 30, 32, 83
cyclooxygenase-2, 2, 16, 30, 32
cyclosporine, 97
cysteine residues, 30
cystine, 8, 15
cytoarchitecture, 119
cytochrome, 11, 12, 16, 21, 32, 33
cytokine response, 69
cytokines, viii, 16, 18, 19, 21, 22, 31, 35, 38, 39, 42, 43, 44, 45, 46, 48, 49, 50, 51, 52, 53, 54, 55, 56, 57, 58, 59, 61, 64, 69, 71, 72, 75, 76, 77, 78, 79, 90, 92
cytopathology, 44
cytoplasm, 21
cytoskeleton, 5, 13, 28
cytotoxicity, 45, 49, 51, 52, 59, 60, 75, 78, 85

D

data analysis, 137
death, vii, 1, 3, 6, 11, 12, 13, 14, 15, 18, 19, 20, 21, 22, 23, 24, 25, 26, 27, 28, 29, 44, 48, 52
decisions, 112, 113, 118
decoupling, 132, 136
defects, 43
defense, viii, 42, 58, 68, 87
defense mechanisms, 42
deficiency, 15, 17, 102
deficit, ix, 76, 107, 110, 120
definition, 3
degradation, 11, 13, 14, 38, 55, 59, 77, 83, 95
delivery, viii, 32, 87, 89, 93, 96, 101
demand, 135
dementia, 17, 36, 37, 40, 60, 64, 67, 71, 74, 78, 80, 82, 93
demyelination, 15
dendrites, 6, 21, 43
dendritic cell (DC), 96
density, ix, 37, 38, 55, 68, 74, 123, 127, 151
depolarization, 27, 32
deposition, 36, 38, 40, 43, 45, 46, 49, 60, 79, 80, 93
deposits, 36, 43, 45, 46, 49, 75
depression, 17, 39, 83, 138
deprivation, 21, 29, 64
deregulation, 53
derivatives, 44
destruction, 5, 19, 63
desynchronization, 125, 129, 134, 139
detection, 36
development, 150, 153
developmental dyslexia, 120
diagnostic criteria, 30
diet, x, 40, 143, 144, 145, 146, 147, 148, 149, 150, 151
dietary intake, 22
differentiation, 63, 118, 150
diffusion, 90
dimorphism, 125, 141

directionality, 19
discs, 74
disease progression, 68
disorder, ix, 17, 25, 36, 66, 107, 110, 120
distribution, 40, 55, 80, 93, 96, 97, 98, 99, 102, 105, 111, 127
divergent thinking, 125, 140
diversity, 111
DNA, 5, 12, 13, 14, 21, 26, 41, 44, 73
DNA breakage, 21
DNA damage, 13, 44
dominance, 109, 110, 111, 112, 113, 114, 116, 119, 120, 121, 152
donors, 8
dopamine, 17, 18, 21, 22, 30
dopaminergic, 17, 30, 85
dopaminergic neurons, 17
dorsolateral prefrontal cortex, 117
down-regulation, 76
drug delivery, viii, 87, 96
drug design, 102
drug discovery, viii, 87
drug efflux, 91, 94, 95, 104
drugs, vii, viii, 1, 22, 25, 39, 41, 64, 82, 87, 91, 93, 94, 95, 98, 101, 102, 104
duration, 50, 55, 81
dyes, 102
dyslexia, 110

E

Education, 141
EEG, ix, 123, 125, 126, 127, 130, 131, 132, 134, 137, 138, 139, 140, 141
EEG patterns, 139
efflux mechanisms, 90, 93, 94, 95
efflux transporters, viii, 88, 92, 93, 94, 95, 104
elderly, 19, 42, 72, 73, 76, 78
elders, 36
electrodes, 127, 130, 131, 132
electroencephalogram, 140
electroencephalograph, 126
electron, 11, 18, 44, 100
electrons, 3, 7, 9
elongation, 19
embryonic development, 46
emission, 46
emotion, 108
emotions, vii, 109, 137, 138
enantiomers, 67
encapsulation, 94
encephalomyelitis, 84
encephalopathy, vii, 1, 6, 24, 25
encoding, 103, 134
endocrine, 152
endogenous, viii, 88, 92
endonuclease, 82
endothelial cells, viii, 31, 68, 87, 88, 89, 90, 91, 92, 96, 97, 99, 100, 102, 103, 104, 105
endothelium, 21, 90, 96, 97, 98
energy, 7, 11, 21, 43
energy supply, 43
environment, viii, 17, 25, 36, 52, 59, 87, 95, 108
environmental conditions, 43, 49
environmental factors, 39
environmental influences, 40
enzymes, viii, 3, 5, 7, 8, 9, 13, 14, 15, 16, 51, 87, 151, 152
ependymal, 90
epidemiology, 72
epididymis, 146
epilepsy, 93, 96, 103, 104, 105, 110, 121
epithelial cells, viii, 87, 88, 89
epithelium, 90, 96, 104
Erk, 74
erythrocytes, 33
estradiol, 92, 144, 146
estrogen, x, 143, 144, 150, 151, 152
ethanol, 28
ethics, 125
ethnicity, 67
ethylene, 2, 20
ethylene glycol, 2, 20
etiology, 93
evidence, 144, 148, 151
evolution, 18, 46, 65, 80, 83, 119
excision, 3
excitability, 43
excitation, 108
excitotoxicity, 9, 13, 15, 18, 19, 22, 49, 62
excretion, 95, 101, 104
exogenous, viii, 88, 92
experimental condition, 48
exposure, 12, 13, 14, 33, 44, 45, 150, 153
externalization, 24
extracellular, viii, 87, 89, 96, 105
extracellular matrix, 50
extraction, 146
extraneous variable, 144
extrusion, 21

F

factor analysis, 146
failure, 21, 45, 80
false positive, 113
family, 17, 53, 55, 56, 57, 62, 65, 92, 97, 99, 101, 102, 103
family history, 17
family members, 57
fat, 97, 146
fatigue, 17
fatty acids, 6
feedback, 7, 50, 58, 59
feet, 91
females, ix, 123, 125, 128, 129, 130, 131, 133, 134, 135, 136
fibroblasts, 65
fixation, 126
flank, 151
fluid, 89, 90, 96, 97, 98, 99, 102, 103, 104, 105
fluid transport, 103
fluorescence, 62, 68
focusing, 18
food, x, 95, 143, 147, 150
forebrain, 85
free radical scavenger, 16
free radicals, 40, 41, 59, 72
friends, 82
frontal cortex, 65, 85, 116, 118, 135
frontal lobe, 108, 135, 136
fuel, 153
functional activation, 117
functional changes, 46, 57, 82
functional imaging, 109, 110, 111, 115, 121
functional MRI, 119, 120, 121, 138

G

gender, 110, 124, 125, 128, 129, 130, 133, 135, 136, 137, 139
gender differences, 124, 125, 130, 133, 135, 136, 139
gene, 13, 45, 66, 74, 76, 91, 96, 97, 100, 101, 102, 103, 104, 105
gene expression, 13, 45, 66, 74, 76, 104
gene targeting, 102
generation, 5, 6, 26, 36, 40, 44, 45, 46, 48, 58, 59, 60, 62, 63, 68, 71, 80, 114, 115, 118, 124, 135, 138, 139
generativity, 119
genes, viii, 12, 13, 36, 38, 66, 80, 81, 85, 97
genetics, vii, 17, 80
genotype, 67
gestures, 108
gifted, 136
glia, viii, 27, 28, 36, 43, 50, 62, 63, 67, 68, 73, 79, 82, 98
glial cells, viii, 33, 35, 38, 39, 40, 42, 43, 44, 45, 47, 49, 50, 51, 52, 53, 54, 55, 56, 59, 60, 61, 62, 66, 75, 76, 82
glucocorticoid, 151
glucocorticoid receptor, 151
glucose, 13, 15, 21, 29
glutamate, 2, 6, 8, 10, 14, 16, 21, 22, 25, 28, 29, 33, 39, 62, 64, 76
glutathione, 2, 3, 4, 7, 8, 9, 18, 25, 27, 28, 61, 101
glutathione peroxidase, 2, 3, 4, 7
glycine, 2, 7, 8
glycolysis, 15
glycoprotein, 90, 96, 97, 98, 99, 100, 101, 102, 103, 104
glycoside, 101
gold, 109
gravity, 127, 132, 133, 134
groups, ix, x, 3, 7, 123, 128, 129, 132, 143, 145, 146, 147, 148, 149, 150
growth, 29, 42, 43, 44, 48, 49, 50, 52, 54, 59, 61, 64, 67, 69, 71, 76, 79, 81, 85, 147
growth factor, 42, 43, 44, 48, 49, 50, 52, 54, 59, 61, 64, 67, 69, 71, 76, 79, 81, 85
growth factors, 43, 49, 50, 52, 54, 59
growth hormone, 43, 71, 76

H

handedness, 110, 125, 138
healing, 71
health, 151
heat, 31
heme, 24, 32, 79, 81
heme oxygenase, 32, 81
hemisphere, ix, 108, 109, 110, 113, 114, 116, 123, 131, 132, 133, 135
hemispheric asymmetry, 110, 111, 129, 141
hepatic, 95
hepatocellular carcinoma, 75
hepatoma, 63
herbicide, 13
heroin, 17

heterogeneity, 38
high density lipoprotein, 68
high scores, 126
hip, 144
hippocampus, 36, 39, 43, 52, 67, 70, 76, 78, 81, 85, 96
histidine, 5
histone, 12, 13
HIV, 93, 99
HIV-1, 99
homeostasis, viii, 6, 13, 40, 46, 50, 51, 53, 65, 88, 90, 92
hormonal control, 144
hormone, 119, 145, 152
hormones, 90, 92, 96
human brain, 6, 38, 56, 70, 79, 81, 83, 84, 91, 102, 103, 108, 111, 119, 140
human immunodeficiency virus, 80
human immunodeficiency virus (HIV), 93, 99
human subjects, 140
hydrocephalus, 90, 93
hydrogen, 2, 3, 4, 9, 10, 29, 30, 45
hydrogen peroxide, 2, 3, 4, 29, 45
hydroperoxides, 4
hydrophobicity, 72
hydroxyl, 2, 3, 4, 5
hyperactivity, ix, 107, 110, 120
hyperbilirubinemia, 24, 27
hyperglycemia, 57
hypoglycemia, 28
hypothalamus, 52
hypothesis, 6, 9, 25, 39, 58, 60, 62, 69, 72, 74, 86
hypoxia, 40, 58, 64

I

ibuprofen, 45
identification, 31, 73, 95, 103, 108
identity, 59
IFN, 44, 46, 47, 48, 51, 52, 53, 54, 68, 71, 72, 75, 84
IL-6, 39, 46, 48, 52, 67
IL-8, 48
imagery, ix, 123, 124, 134, 136, 137, 138, 139, 140
images, ix, 115, 123, 124, 125, 126, 128, 131, 135
imaging, 104, 109, 121, 138
imaging techniques, 109
imbalances, 19, 70
immune function, 44
immune response, 76
immune system, 46
immunity, 46, 76
immunization, 40, 76
immunocompetent cells, 46, 81
immunoreactivity, 61, 71, 79
immunotherapy, 23, 68
impairments, 76
implementation, 109
in situ, 84
in vitro, 13, 28, 38, 41, 45, 47, 49, 53, 56, 64, 67, 69, 73, 78, 81, 82, 83, 84, 90, 91, 99, 100, 104, 152
in vivo, 13, 32, 38, 41, 45, 47, 53, 56, 60, 67, 83, 90, 152
incidence, vii, 17, 20, 30, 35, 40, 110
inclusion, 116
independent variable, 150
indication, 78
indicators, 117
indices, ix, 115, 123, 135
individual differences, 140
inducer, 3, 4
induction, 7, 27, 44, 48, 51, 53, 54, 56, 57, 58, 59, 68, 70, 77, 82, 83, 84
industry, viii, 87, 93
inefficiency, 5
infants, 141
infarction, 32
infection, 27, 46, 51
infections, 91, 93
inferior vena cava, 146
inflammation, vii, 16, 18, 35, 39, 41, 43, 44, 45, 46, 51, 52, 55, 56, 57, 58, 59, 61, 63, 70, 73, 77, 79, 82
inflammatory, 105
inflammatory mediators, 40, 49, 57
inflammatory response, vii, 21, 31, 35, 36, 38, 39, 40, 44, 45, 46, 48, 49, 50, 52, 53, 54, 55, 56, 57, 59, 60, 69, 71, 75, 77, 82, 105
inflammatory responses, 21, 36, 39, 49
informed consent, 125
inhibition, 5, 7, 11, 15, 16, 17, 18, 31, 51, 55, 71, 82, 93, 94, 101, 134
inhibitor, 14, 32, 98
inhibitory effect, 50, 105
injuries, 40, 49, 52
input, 74
insight, 93
instruction, ix, 123, 125, 128, 129, 130, 131, 132, 133, 134, 135, 136
insulin, 43, 55, 71, 76, 81, 84
integration, 6, 52, 135, 137, 141

integrin, 63, 74, 84
integrity, 58, 59, 135
intellect, 111
intensity, 115, 116
interaction, 3, 4, 20, 21, 30, 50, 51, 53, 57, 60, 66, 70, 75, 82, 128, 129, 139
interaction effect, 66
interaction effects, 66
interactions, 50, 60, 68, 72, 78, 127, 128, 129, 136, 144
interest, 150
interface, 101
interferon, 46, 62, 63, 69, 71, 75, 77, 79
interferon gamma, 46, 62
Interleukin-1, 62, 65, 68, 72, 76, 78
interleukins, 44
intermediaries, 50, 59
internalization, 55
interpretation, vii, 19, 39, 109
interval, 126, 151
intervention, 73
intestine, 100
intonation, 121
intravenous, 104
ion channels, 77
ion transport, 92
ions, 9, 10, 20, 21, 22
iron, 3, 4, 5, 15, 18, 19, 20, 30
ischemia, 7, 21, 22, 26, 28, 40, 49, 53, 69, 85
ischemic stroke, 72
isolation, 111
isomerization, 7

K

K^+, 10, 13, 24
kidney, 95, 99, 100, 101
kinase C inhibitors, 94
kinetics, 54, 105

L

labeling, 62
language, ix, 107, 108, 109, 110, 111, 112, 115, 119, 120, 121, 140, 141
language lateralization, ix, 107, 109, 110, 112, 120
language processing, 140
laterality, ix, 107, 110, 111, 112, 113, 115, 116, 117, 118, 134
LDL, 55, 72, 78
lean body mass, 83
learning, vii, 64, 139
left hemisphere, ix, 107, 108, 110, 114, 116, 123, 124, 129, 131, 132, 133, 134, 135
left-handers, 110
left-hemisphere, 113, 114
leptin, 144, 150
Leptin, 153
lesions, ix, 36, 40, 44, 63, 66, 81, 107, 108, 109, 110, 120
leucine, 2, 10, 11
levodopa, 30
ligand, 57, 78
ligands, 57, 72, 98
lignans, 144
likelihood, 40
lipid, 102
lipid peroxidation, 3, 4, 5, 6, 11, 13, 14, 19, 21, 24, 26, 29, 44, 64, 74
lipids, 3, 4, 18, 33, 41, 44, 55
lipoproteins, 68
liver, 95, 96, 100, 105
localization, 77, 84, 91, 100, 102, 103, 104
location, 104
low-density lipoprotein, 84
low-density lipoprotein receptor, 84
LPS, 33, 44, 45, 46, 47, 48, 51, 53, 54, 67, 69, 72, 73, 84
lung cancer, 105
lysine, 5

M

machinery, 6, 16
macrophages, 46, 51, 54, 55, 56, 72, 74, 80, 84
magnetic resonance, 109, 119, 140
magnetic resonance imaging, 109, 119, 140
magnetoencephalography, 109
males, ix, x, 123, 125, 128, 129, 130, 131, 133, 134, 135, 136, 143, 150
malignant, 92
mammalian brain, 68, 76
management, 33
manipulation, x, 143, 145, 146, 147, 148, 149, 150
mapping, 119, 131
MARCO, 55, 56, 57
Marx, 63
masking, 117, 118
mathematics, 139

matrix, 11
MCP, 43, 45, 48, 49, 75
MCP-1, 43, 45, 48, 49, 75
mean, 148
measurement, 64, 91, 140
measures, viii, 87, 112, 120, 125, 128, 129, 133, 134, 138
media, 48, 51
mediation, 96, 137
medication, 93
MEG, 109
membrane permeability, 11, 12
membranes, 11, 19, 71, 72, 81, 91, 97
memory, vii, 6, 19, 39, 40, 43, 62, 81, 110, 111, 119, 124, 134, 135, 139, 140, 141
memory processes, 62, 135
men, ix, 17, 61, 124, 125, 129, 130, 131, 132, 135, 137, 139
meningitis, 94
mental arithmetic, 139
mental image, 124, 135, 138, 139
mental imagery, 124, 138, 139
mental processes, 111, 124, 134
mesangial cells, 69
messenger RNA, 64, 103
messenger RNA (mRNA), 95, 103
messengers, 42
meta-analysis, 67, 121, 139, 141
metabolic changes, viii, 36, 40
metabolic pathways, 28
metabolism, 9, 10, 11, 15, 23, 25, 28, 31, 51, 55, 68, 92, 103
metabolites, 17, 18, 45
metabolizing, 151
metals, 18, 30, 36
methionine, 93
mice, 13, 17, 30, 32, 38, 40, 43, 45, 49, 53, 56, 61, 63, 64, 66, 69, 71, 75, 77, 80, 82, 83, 84, 85, 91, 95, 98, 101, 102, 103, 105, 144, 150, 152, 153
microcirculation, 40
microdialysis, 78, 97, 98, 105
microglia, 16, 18, 21, 22, 23, 24, 25, 30, 33, 36, 40, 42, 43, 44, 45, 46, 47, 48, 49, 50, 51, 53, 56, 57, 59, 61, 63, 64, 65, 67, 68, 69, 70, 71, 72, 73, 75, 77, 78, 80, 81, 82, 84
microinjection, 39
microscopy, 101
microstructure, 111
migration, 57
Ministry of Education, 118
MIP, 48
mitochondria, 11, 18, 21, 44, 64, 85
mitochondrial damage, 16, 18
mitochondrial DNA, 29
mitochondrial toxins, 15
mitogen, 30, 57, 69, 75, 83, 86
models, 16, 18, 26, 44, 49, 52, 62, 110, 111, 152
molecular mechanisms, 60
molecular oxygen, 3
molecular pathology, 32
molecular weight, 3
molecule, 94
molecules, 3, 4, 40, 41, 43, 44, 45, 46, 47, 48, 50, 51, 52, 53, 54, 55, 56, 57, 59, 70, 78, 94
momentum, 95
monoclonal antibody, 103
monoclonal antibody (mAb), 103
monocyte chemoattractant protein, 43
monocytes, 61, 65, 75
monomers, 15, 85
mononuclear cells, 53
mood, 111
morbidity, 43
morphological abnormalities, 42, 43
morphology, x, 43, 63, 74, 143, 150
morphometric, 65
mouse model, 30, 62, 73, 74, 78
movement, 9, 10, 17, 127
MRI, 109
mRNA, 52, 53, 68, 78, 84, 95
mucosa, 101
multiple factors, 59
multiple sclerosis, 70, 81
multiple sclerosis (MS), 97, 102
muscle cells, 62
music, 108
musicians, 121
mutagenesis, 13
mutations, 5, 38
myelin, 6, 57, 74
myelinization, 15

N

Na^+, 10, 13, 24
nanoparticles, 94, 100
natural science, 136
natural sciences, 136
necrosis, 21, 24, 29, 39, 62, 65, 81
neglect, 111

neocortex, 7, 27
neonates, 24
neoplasia, 144, 150, 152
nerve, 1, 2, 6, 16, 21, 22, 27, 28, 29, 37, 48, 64, 70, 72, 79, 84
nerve cells, 1, 21, 27, 37
nerve growth factor, 2, 16, 48, 64, 70
nervous system, 6, 15, 29, 41, 44, 50, 68, 70, 71, 82, 96, 98, 100, 101, 103, 151
network, 6, 117
neural network, 137
neural tissue, 6
neurobiology, 71, 138
neurodegeneration, viii, 19, 21, 22, 23, 25, 26, 29, 36, 41, 45, 48, 58, 62, 66, 69, 75, 76, 77, 84
neurodegenerative diseases, vii, 35, 39, 41, 43, 44, 49, 50, 51, 60, 61, 73, 75, 151
neurodegenerative disorders, vii, 1, 16, 17, 25, 28, 29, 42, 74, 96
neurodegenerative processes, 23, 58, 59
neuroendocrine, 151, 152
neurofibrillary tangles, 36, 37, 58
neurohormone, 7
neuroimaging, 111, 112, 113, 118, 120, 121
neuroinflammation, 44, 45, 62, 80
neurological deficit, 32
neurological disease, vii, 1, 6, 52
neurological disorder, 93, 100
neuronal apoptosis, 26, 65
neuronal cells, 38, 51, 63, 81, 83
neuronal death, 10, 16, 18, 21, 26, 64, 73, 80
neuronal plasticity, 29
neuronal sensitivity, 45
neuronal survival, 9, 29, 39, 50, 68
neuronal systems, 111
neurons, vii, 1, 6, 7, 8, 9, 10, 11, 12, 13, 14, 15, 16, 17, 19, 21, 22, 24, 25, 26, 28, 29, 32, 38, 43, 44, 45, 46, 48, 49, 50, 51, 52, 53, 54, 55, 57, 64, 65, 66, 67, 69, 74, 76, 77, 80, 81, 82, 83, 84, 85, 97, 108, 120, 138
neuropeptide, 27
neuropeptides, 85
neuroprotection, 18, 26, 53, 64, 66, 72, 76
neuroprotective, 18, 30, 38, 39, 49, 52, 53, 59, 80, 81
neuroscience, 110, 139
neurotoxic effect, 85
neurotoxicity, 19, 26, 28, 38, 48, 53, 55, 57, 58, 59, 61, 62, 67, 69, 73, 77, 81, 83, 85, 94
neurotransmitter, 10, 24, 25, 39
neurotransmitters, 21, 22, 28, 39, 52, 90, 92, 97
neurotrophic factors, 16, 48, 53
nicotinamide, 7, 9
nitric oxide, viii, 2, 3, 4, 13, 21, 29, 36, 61, 62, 63, 65, 68, 69, 70, 71, 72, 74, 77, 82, 83
nitric oxide synthase, 2, 13, 21, 29, 65, 69, 74, 77, 82, 83
nitrogen, 29, 50
nitrosative stress, viii, 36
nodes, 127, 131
non-human primates, 104
non-steroidal anti-inflammatory drugs, 45, 67
norepinephrine, 39
NSAIDs, 67, 83
nucleic acid, 3, 5, 19, 21, 22
nucleus, 13, 14, 52, 138

O

object recognition, 136
observations, 39, 45, 56, 57, 60, 91, 93
occipital cortex, 118
occipital regions, 136
occlusion, 32, 85
olanzapine, 119
old age, 36, 42, 63
oligodendrocytes, 6, 15, 16, 29, 30, 38, 61
oligomers, 56, 71, 83
opiates, 93
optic nerve, 72, 78, 79
optimization, viii, 87
organ, vii, 1
organelles, 21
organism, 16
organization, 111, 112, 125, 137, 140, 141
orientation, 108, 109, 121
originality, ix, 123, 126, 128, 136, 140
overproduction, 38, 59
ovulation, 150
oxidation, 3, 17, 18, 19, 21, 44, 73, 153
oxidative damage, vii, 1, 3, 5, 9, 13, 14, 16, 23, 24, 25, 39, 86
oxidative stress, vii, 1, 3, 5, 6, 7, 9, 11, 12, 13, 14, 15, 16, 17, 18, 19, 20, 21, 22, 23, 24, 25, 26, 27, 28, 29, 30, 31, 32, 37, 41, 45, 46, 49, 55, 57, 58, 60, 64, 68, 79, 80, 83
oxygen, 3, 4, 5, 8, 21, 40, 42, 51, 109
oxygen consumption, 42

P

pain, 93, 152
parenchyma, 40, 89
parietal cortex, 118, 136, 141
parietal lobe, 137
Parkinson's disease, 93
parkinsonism, 17, 18, 30
parole, 108
partition, viii, 87
passive, 40, 90
pathogenesis, 6, 17, 20, 21, 38, 39, 46, 52, 58, 66, 71, 75, 76, 80, 96, 100
pathogens, 71
pathological aging, 42, 60
pathology, 38, 39, 40, 43, 48, 52, 60, 61, 62, 73, 74, 75, 78, 79, 82, 83, 96, 110
pathophysiological mechanisms, 40, 44
pathways, vii, 1, 6, 11, 17, 18, 20, 22, 23, 24, 25, 30, 51, 52, 54, 55, 56, 57, 59, 60, 75, 84, 140
penicillin, 90, 93, 94, 98, 103
penicillins, 94
peptide, 93, 102
peptides, 9, 38, 39, 62, 63, 67, 69, 75, 84, 90, 92, 94
perception, 6, 119, 137
perfusion, 40, 90
peripheral blood, 75
peripheral blood mononuclear cell, 75
permeability, 11, 96, 102
peroxidation, 4, 5, 11, 21, 23, 25, 26, 41, 44
peroxide, 2, 3, 30, 62
peroxynitrite, 3, 4, 28, 44, 85
PET, 93, 109, 135, 136, 139, 140
pH, 33, 103
phagocytosis, 24, 46, 57, 66, 74, 78, 83
pharmacokinetics, 97, 98, 102
pharmacotherapy, 69
phosphatidylserine, 24, 33
phosphorylation, 3, 4, 5, 11, 13, 14, 17, 18, 26, 54, 85
phrenology, 111
physiology, 145, 150
pilot study, 117
placenta, 99, 105
planning, 135
plaque, 36, 38, 52, 61, 65, 66, 73, 80, 83, 84
plasma, 11, 13, 17, 21, 24, 40, 55, 97, 105, 151
plasma membrane, 11, 13, 17, 21, 24, 40, 55
plasminogen, 82
plasticity, 29, 39, 51, 62, 81, 119
plexus, viii, 87, 88, 89, 90, 91, 92, 95, 96, 97, 101, 102, 104
PM, 96, 98, 119, 137
polymerase, 21, 84
polymerase chain reaction, 84
polymorphism, 103
polymorphisms, 93
polypeptide, 96
polypeptides, 70
polyphenols, 29
polyunsaturated fat, 4
polyunsaturated fatty acids, 4
poor, viii, 87, 93
population, 24, 42, 83, 109, 118
Portugal, 1
positive feedback, 57
positive relation, 144
positive relationship, 144
positron, 109, 139
positron emission tomography, 109, 139
post-transcriptional regulation, 99
potassium, 6, 9, 10, 17, 24, 90, 99
power, ix, 123, 125, 127, 128, 129, 134
precursor cells, 15
prediction, 119
preference, 137
prevention, 28, 101
primary tumor, 104
primary visual cortex, 136, 138
priming, viii, 36, 42, 59
prion diseases, vii, 1, 22, 23, 25
production, vii, 1, 3, 11, 13, 14, 15, 16, 17, 18, 19, 21, 22, 24, 25, 33, 37, 38, 39, 40, 41, 43, 45, 46, 47, 48, 49, 50, 51, 52, 53, 54, 55, 56, 57, 58, 59, 60, 61, 62, 63, 64, 65, 70, 72, 75, 77, 78, 80, 81, 82, 147
progesterone, 146
program, 78, 126
progressive neurodegenerative disorder, viii, 36
pro-inflammatory, viii, 18, 35, 36, 38, 39, 44, 45, 46, 47, 48, 49, 50, 51, 52, 53, 54, 55, 56, 57, 58, 59, 60, 72, 78, 84
pro-inflammatory response, 55
proliferating fraction (PF), 98
proliferation, 16, 43, 57, 61, 72
proposition, 39
prostaglandins, 18, 19, 52
protease, 99
protease inhibitor, 99
protease inhibitors, 99

proteases, 11, 12, 38
protective role, 7, 39, 40, 51, 52
protein, 91, 92, 94, 95, 96, 97, 98, 99, 100, 101, 102, 103, 104, 105
protein family, 97
protein function, 13
protein kinase C, 27, 94
protein kinases, 78
protein oxidation, 19, 21, 24, 26
protein structure, 5
protein synthesis, 5
proteins, viii, 7, 11, 13, 15, 18, 19, 21, 22, 27, 31, 36, 41, 43, 44, 49, 53, 54, 55, 56, 62, 65, 70, 75, 77, 80, 83, 92, 94, 96, 101, 103, 151, 152
proteoglycans, 55
proteolysis, 22
proteolytic enzyme, 59
PSD, 68
psychology, vii, 139
pumps, 19
purification, 77
purines, 13

Q

quality of life, 25
quaternary ammonium, 102
query, 111
quinolone antibiotic, 101
quinone, 18

R

RA, 96, 98, 101, 102
rain, 29, 91
range, viii, 35, 91, 94, 113, 131, 132, 136, 141
RANTES, 45, 49
reaction time, 137
reactive gliosis, 49, 75
reactive oxygen, vii, 1, 3, 4, 12, 14, 18, 20, 22, 23, 24, 28, 29, 39, 65, 71
Reactive Oxygen Species, 15, 16
reactivity, 38, 40, 42, 43, 47, 52, 58, 59, 69, 82, 97, 125, 127, 128, 129, 130, 131, 132
reading, 60
reception, vii, 6
receptor, 98
receptors, 5, 7, 13, 22, 24, 25, 52, 56, 57, 59, 60, 63, 64, 70, 71, 72, 74, 94, 144, 150
recognition, 6, 24, 63, 84, 108
recovery, 50
recycling, 8
redistribution, 17, 18
reduction, 7, 13, 16, 43, 69, 71, 145
refractory, 93
regeneration, 15, 16, 61
regulation, vii, 23, 27, 28, 39, 43, 46, 50, 55, 59, 70, 76, 77, 79, 82, 95, 99, 135
regulators, 11
relationship, 94, 110, 134, 144
relationships, 110, 134
relevance, 7, 30, 63, 66, 100, 101, 120
reliability, ix, 107, 112, 113, 121
remission, 42
remodelling, 12, 13
remyelination, 61
repair, 21, 40, 43, 53, 73
reparation, 3
replication, 12, 13
repressor, 13, 14
resistance, 9, 13, 16, 29, 74, 90, 91, 92, 96, 97, 98, 99, 100, 101, 102, 103, 104, 105, 126
resolution, 137
resources, 140
respiration, 15, 25, 44, 62
respiratory, 11, 28, 30, 48, 56
retrovirus, 68
returns, 46
rhythm, 108, 121, 125
ribose, 21
ribozymes, 94
right hemisphere, ix, 107, 108, 109, 114, 124, 129, 131, 132, 133, 134, 135, 139
right-handers, 110
risk, 15, 19, 21, 24, 38, 40, 41, 42, 44, 67, 68, 71, 78, 82, 109, 110, 127, 152
risk factors, 15, 38
RNA, 5, 75, 103
robustness, 112, 115, 120
ROOH, 7
room temperature, 146
RS, 99

S

sacrifice, 147
safety, 91
sample, 19, 126, 146
sampling, 126

schizophrenia, ix, 107, 110, 111, 119, 121, 139
school, 133
sclerosis, 119
scores, ix, 123, 124, 125, 126, 127, 128, 130, 133
search, 110
secrete, 42, 43, 45, 48, 49, 50, 53, 56, 57
secretion, 24, 25, 42, 53, 54, 56, 57, 59, 63, 67, 68
seed, 71
selective estrogen receptor modulator, 150, 152
semantic processing, 136
senile dementia, 63
senile plaques, 19, 36, 43, 49, 56, 61, 65, 66, 72, 74
sensitivity, 45, 94, 101, 102, 104
septum, 150
sequencing, 22, 102
series, 54, 59
serine, 8, 54, 84
SERMs, 150, 151, 152
serum, 146, 147, 148, 150
severity, 37, 65, 79
sex, ix, 30, 67, 83, 123, 124, 125, 133, 134, 135, 137, 139, 140, 141, 150, 152
sex differences, ix, 123, 124, 125, 133, 134, 135, 137, 139, 141
sex hormones, 83, 152
shape, 111
shares, 39
shock, 31
short-term memory, 137
sign, 113, 116
signal transduction, 52, 55, 56, 57, 66, 67, 72
signaling pathway, 53, 56, 60, 66, 69, 70, 72, 75, 86
signalling, 25, 28, 74, 79
signals, 6, 52, 59, 60, 72
signs, 65, 128, 129, 130
silver, 36
sites, ix, 5, 27, 49, 57, 84, 97, 109, 123, 126, 131, 132, 133
skills, 124, 138
skin, 71, 146
sleep disturbance, 17
smooth muscle, 21, 70
smooth muscle cells, 21, 70
social behavior, 144, 150
sodium, 6, 7, 8, 9, 10, 17, 21, 24, 27, 103, 121, 146
solubility, 3
solution, 94, 146
spatial ability, 124, 134, 139
spatial processing, 141
specialization, 6, 108, 111
species, vii, 1, 3, 4, 8, 12, 14, 18, 21, 22, 23, 24, 28, 29, 39, 54, 58, 65, 71, 72, 91, 99
specificity, 94, 95
speculation, 150
speech, 6, 108, 119, 121
speech perception, 119
spinal cord, 26, 27, 77, 103
spinal cord injury, 77
sprouting, 67
stages, 38, 45, 46, 51, 52, 58, 79
standard error, 148
steroid, 96
steroids, 76
stimulus, 42, 45, 54, 60, 134
storage, 7
strain, 5
strategies, viii, 62, 64, 73, 87, 94, 124, 137
stress, vii, 1, 3, 5, 6, 9, 13, 15, 17, 18, 19, 20, 22, 23, 24, 25, 26, 28, 29, 31, 41, 44, 55, 59, 60, 76, 77, 81, 151
stroke, vii, 1, 6, 15, 21, 22, 25, 31, 82
structural changes, 41
students, ix, 60, 123, 124, 125, 126, 133, 135, 141
subcutaneous (SC), 96, 105
substrates, 7, 8, 9, 12, 92, 93, 94
subtraction, 127
suffering, 38
sulfate, 96
superiority, ix, 107
supply, 8
suppression, 43
surprise, 16
surveillance, 6
survival, 3, 24, 29, 31, 50, 52, 53, 84, 109
susceptibility, ix, 14, 15, 28, 40, 59, 68, 85, 107, 110, 120
suture, 146
swelling, 21
symptom, 17
symptoms, 109
synapse, 19, 48
synaptic plasticity, 40, 43, 64, 66
synaptic transmission, 39, 81
synaptic vesicles, 10, 17, 18
synchronization, 137, 138, 139, 141
syndrome, 17, 24
synergistic effect, 47, 83
synthesis, viii, 7, 8, 9, 10, 36, 45, 48, 66, 68, 73
systems, 3, 50, 139

T

T cell, 96
tangles, 36, 37
targets, 6, 13, 14, 15, 21, 28, 52
task demands, 127
task difficulty, 134
task performance, 126, 127, 130, 136
technology, 125
temperature, 146
temporal lobe, 96, 119, 120
temporal lobe epilepsy, 96, 119, 120
terminals, 21, 22, 28, 68, 84
territory, 150
test scores, 138
testes, x, 143, 147, 148, 150
testicle, 146
testis, 99
testosterone, x, 143, 144, 145, 146, 147, 148, 149, 150, 153
testosterone levels, x, 143, 144, 145, 147, 149, 150
TGF, 44, 45, 46, 49, 50, 51, 52, 53, 54, 59, 64, 66, 69, 71, 72, 75, 85
theory, viii, 20, 22, 23, 36, 119, 121
therapeutic approaches, vii, 1, 6, 17
therapeutic targets, vii, 1, 21, 25
therapeutics, 69
therapy, 18, 75, 80
thinking, ix, 17, 123, 124, 125, 135, 139, 140
threat, 3, 4, 21
threshold, 9, 10, 24, 63, 113, 114, 115, 116
thresholds, 113, 114, 115
thyroid, 96
time, 7, 15, 17, 46, 52, 54, 59, 95, 109, 112, 119, 124, 132, 147
timing, 55, 79
tissue, 26, 37, 50, 52, 53, 57, 72, 82, 91, 92, 96, 97, 102, 105, 115, 146
tissues, 91, 92, 97, 99, 100, 104
TNF, 3, 16, 39, 44, 45, 46, 48, 49, 51, 52, 53, 55, 57, 61, 66, 67
TNF-alpha, 66, 67
toxicity, 19, 24, 29, 30, 33, 43, 48, 49, 62, 64, 65, 73, 74, 79
transcranial Doppler sonography, 109, 119, 120
transcription, 15, 23, 37, 54, 57
transcription factors, 23
transduction, 54, 56
transection, 79
transformation, 43, 127
transforming growth factor, 64, 66, 72, 74, 75, 77, 82, 85, 86
transition, 9, 11, 32, 80
transition metal, 9, 32
translocation, 21, 52
transmission, 22, 52
transport, viii, 6, 7, 9, 15, 28, 29, 44, 71, 84, 87, 90, 94, 95, 96, 97, 98, 99, 100, 101, 102, 103, 104, 105
transport processes, 96
trauma, 70
traumatic brain injury, 52
treatment, 93, 94, 100, 101
tremor, 17
trend, 131
trial, 18, 68
triggers, 47, 56
trophic support, 48
Tukey HSD (post hoc), 146
tumor, 3, 16, 45, 54, 69, 78, 93, 97, 150
tumor necrosis factor, 3, 16, 45, 69, 78
tumors, 100, 104
twins, 30
type II error, 127
tyrosine, 51, 57

U

unconjugated bilirubin, 3, 24, 33
university students, 133
urine, 90

V

validation, 110
validity, 140
values, 113, 127, 128, 129, 130, 151
variability, 110, 113, 115, 119, 146
variable, 112, 144, 150
variables, 127
variance, 146
variation, 114, 140
vascular dementia, 40
vasopressin, 150
vector, viii, 87, 102
ventricles, 90
verbal fluency, 121
vertebrates, 91
vesicle, 84, 85

vessels, 97, 146
vincristine, 94
visual images, ix, 123, 134, 135
visual perception, 139
visual processing, 136
vitamin C, 31
vitamin E, 13, 20, 24, 76
vulnerability, 15

W

water, x, 143, 145, 147
web, 85
women, ix, 111, 124, 125, 129, 130, 131, 132, 135, 137, 139
work, 151
workers, 90
working memory, 137, 138, 139, 140
writing, 95

Y

yield, 8, 112

Z

zinc, 21, 22, 31